# QUANTUM MECHANICS
### USING
# COMPUTER ALGEBRA

# QUANTUM MECHANICS
## USING
# COMPUTER ALGEBRA

Includes sample programs for
REDUCE, MAPLE, MATHEMATICA and C++

**Willi-Hans Steeb**

Rand Afrikaans University
Johannesburg

**World Scientific**
Singapore • New Jersey • London • Hong Kong

*Published by*

World Scientific Publishing Co. Pte. Ltd.

P O Box 128, Farrer Road, Singapore 9128

*USA office:* Suite 1B, 1060 Main Street, River Edge, NJ 07661

*UK office:* 73 Lynton Mead, Totteridge, London N20 8DH

**QUANTUM MECHANICS USING COMPUTER ALGEBRA**

ISBN 981-02-1770-6

Printed in Singapore by Utopia Press.

# Contents

Comment on the programs                                                       1
1. Conservation law and Schrödinger equation                                  4
2. Wave-packet and free Schrödinger equation                                  6
3. Separation ansatz and Schrödinger equation                                 8
4. Matrix representation in the Hilbert space $L_2(-\pi, \pi)$               10
5. One-dimensional potential and trial function                             12
6. Heisenberg equation of motion                                            14
7. Harmonic oscillator                                                      18
8. Harmonic oscillator and recursion relation                               20
9. Commutation relations of $\hat{p}$ and $\hat{q}$                         22
10. Anharmonic oscillator                                                   24
11. One-dimensional WKB-solutions                                           26
12. Angular momentum operators I                                            28
13. Angular momentum operators II                                           30
14. Angular momentum operators III                                          32
15. Lie algebra $su(3)$ and commutation relations                           34
16. Spin-1 Lie algebra and commutation relations                            36
17. Radial symmetric potential and bound states                             38
18. Wave function of hydrogen atom I                                         42
19. Wave function of hydrogen atom II                                        44
20. Helium atom and trial function                                          48
21. Stark effect                                                            52
22. Scattering in one dimension                                             58
23. Gauge theory                                                            62
24. Driven two level system                                                 64
25. Free electron spin resonance                                            66
26. Two point Ising model with external field                               70
27. Two point Heisenberg model                                              74
28. Fermi operators                                                         78
29. Fermi operators with spin and the Hubbard model                         82
30. Bose operators                                                          86
31. Matrix representation of Bose operators                                 88
32. Coherent states                                                         90
33. Quartic Hamilton operator and Bose operators                            92
34. Dirac equation and dispersion law                                       94
35. Perturbation theory                                                     98

36. Elastic scattering ................................................ 104
37. Exceptional points ............................................... 108
38. Expansion of $\exp(L)A\exp(-L)$ ............................ 110
39. Expansion of $(A - \epsilon B)^{-1}$ ........................... 114
40. Heavyside, sign-, and delta function ....................... 118
41. Legendre polynomials .......................................... 120
42. Laguerre polynomials .......................................... 122
43. Hermite polynomials ........................................... 124
44. Chebyshev polynomials ........................................ 126
45. Spherical harmonics ............................................ 128
46. Clebsch-Gordon series .......................................... 130
47. Hypergeometric functions ..................................... 132
48. Eigenvalue problems and hypergeometric differential equation ... 136
49. Gamma matrices and spin matrices .......................... 140
50. Fourier transform ............................................... 142
51. Discrete Fourier transform .................................... 146
52. Fourier expansion .............................................. 148
53. Group theory ................................................... 150
54. Quantum groups ............................................... 154
55. Gram-Schmidt orthogonalisation process ................... 158
56. Soliton theory and quantum mechanics ..................... 160
57. Pade approximation ............................................ 166
58. Cumulant expansion ........................................... 170
59. Leverrier's method ............................................. 172
Mathematica programs ............................................. 174
Maple programs ................................................... 176
C++ programs ..................................................... 179
References ......................................................... 185
Index ............................................................... 187

# Preface

Solutions to problems in quantum mechanics are important for scientists, engineers and many others. This book gives a collection of most standard methods in quantum mechanics together with their programs in REDUCE. The output of the programs is also displayed. All the problems in the book are implemented in REDUCE. For a number of selected problems we also give the implementation in MATHEMATICA, MAPLE and C++.

REDUCE, MATHEMATICA and MAPLE are the most widely available and simple to use computer algebra systems. They enable users to manipulate algebraic expressions and equations symbolically. For example, we can differentiate and integrate symbolically. Number crunching can also be done. Moreover, symbolic manipulation and number crunching can be combined in one program.

Beside the standard methods, modern developments in quantum mechanics are also included. These include Bose operators, Fermi operators, coherent states, gauge theory and quantum groups. All the special functions (such as Hermite, Chebyshev, Legendre) are implemented.

The level of presentation is such that one can study the subject early on in ones education in science. There is a balance between practical computation and the underlying mathematical theory. The book is ideally suited for use in a quantum mechanics lecture.

The reference list gives a collection of text books useful in the study of the computer language REDUCE, MATHEMATICA and MAPLE.

All programs have been run on a 486 AT computer under DOS 5.0 using the REDUCE version 3.4.1 and 3.5. The MAPLE programs have been run on version V. The MATHEMATICA programs have been run on version 2.1 for Windows. The C++ programs have been run under Borland C++ (version 3.0) and Turbo C++ (version 3.0) for DOS. The programs also run under Borland C++ (version 1) for OS/2 (version 2.1).

Without doubt, this book can be extended. If you have comments or suggestions, we would be pleased to have them. The email address of the author is:
WHS@RAU3.RAU.AC.ZA

REDUCE is a registered trademark of The RAND corporation. MATHEMATICA is a registered trademark of Wolfram Research. MAPLE is a registered trademark of Waterloo Maple Software. DOS and Microsoft Windows are registered trademark of Microsoft. Borland C++ and Turbo C++ are registered trademarks of Borland International.

# Symbol Index

Here we give a list of symbols used in the text and in the programs. The symbol on the left-hand side gives the symbol used in the text and the symbol on the right hand side that used in the REDUCE programs.

| | |
|---|---|
| $\psi \to psi$ | wave function, solution of the Schrödinger equation |
| $\psi^* \to psis$ | complex conjugate wave function |
| $\rho \to rho$ | density $\rho = \psi^* \psi$ |
| $u \to u$ | solution of the eigenvalue equation |
| $i = \sqrt{-1} \to i$ | $\sqrt{-1}$ |
| $m \to m$ | mass |
| $\hbar \to hb$ | Planck constant divided by $2\pi$ |
| $\hat{H} \to H$ | Hamilton operator |
| $s_x, s_y, s_z \to sx, sy, sz$ | Pauli spin matrices |
| $\hat{L}_x, \hat{L}_y, \hat{L}z \to Lx, Ly, Lz$ | angular momentum |
| $\hat{p} \to p$ | momentum operator |
| $\hat{q} \to q$ | coordinate |
| $E \to en$ | energy |
| $V \to V$ | potential |
| $\otimes \to kron$ | Kronecker product |
| $\mathrm{tr} \to trace$ | trace |
| $\det \to det$ | determinant |
| $I \to ID$ | unit matrix, identity operator |
| $\delta_{jk} \to delta$ | Kronecker delta |
| | $\delta_{jk} = 1$ for $j = k$ and $\delta_{jk} = 0$ for $j \neq k$ |
| $\lambda \to lam$ | eigenvalue |

# Quantum Mechanics
# Using
# Computer Algebra

by
Willi-Hans Steeb
Rand Afrikaans University, Johannesburg, South Africa

in collaboration with

Dirk Lewien
Technische Hochschule Darmstadt, Darmstadt, Germany

Choy-Heng Lai
National University of Singapore, Singapore

W. D. Heiss
University of the Witwatersrand, Johannesburg, South Africa

Catharine Thompson
Rand Afrikaans University, Johannesburg, South Africa

## Comment on the Programs

We now give a summary of the commands in REDUCE most commonly used in our programs. Then we give a list of commands we used most in the MATHEMATICA and MAPLE programs. For a complete summary of commands we refer to the user manuals (see reference list).

First we remark that REDUCE does not distinguish between capital and small letters. Thus the commands $sin(x)$, $Sin(x)$ or $SIN(X)$ are the same.

The commands we use most in this book are differentiation and integration. To differentiate $x^3 + 2x$ with respect to $x$ we write

$$df(x**3 + 2*x, x)$$

The output is $3x**2 + 2$. To integrate $x^2 + 1$ we write

$$int(x**2 + 1, x)$$

The output is $x**3/3 + x$.

The command $solve()$ solves a number of algebraic equations, and also systems of algebraic equations. For example, the command

$$solve(x**2 + (a + 1)*x + a = 0, x)$$

gives the solution $x = -1$ and $x = -a$.

Another important command is the substitution command, $sub()$. For example, the command

$$sub(x = 2, x*y + x**2)$$

yields $2*y + 4$.

Amongst others, REDUCE includes the following mathematical functions: $sqrt(x)$ (square root, $\sqrt{x}$), $exp(x)$ (exponential function, $e^x$), $log(x)$ (natural logarithm $\ln(x)$, and the trigonometric functions $\sin(x)$, $\cos(x)$, $\tan(x)$ with arguments in radians.

REDUCE reserves $i$ (or $I$) to represent $\sqrt{-1}$ and $pi$ (or $PI$, $pI$, $Pi$) for the number $\pi$. Thus the input $i*i$ gives $-1$ and $\sin(pi) = 0$. Other predefined constants are $E$, $Infinity$, $nil$, $T$. $T$ stands for true.

For other commands we refer to the user manual for REDUCE (see references).

MATHEMATICA distinguishes between capital letters and small letters. The command sin[0.1] for the evaluation of the sine of 0.1 gives the error message, "possible spelling error", whereas Sin[0.1] gives the right answer 0.0998334. .

In MATHEMATICA the differentiation command

$$D[x^\wedge 3 + 2*x, x]$$

gives the output $2 + 3x^2$. The command

$$Integrate[x^\wedge 2 + 1, x]$$

gives $x^3/3 + x$.

The command *solve*[ ] can solve a number of algebraic equation and also systems of algebraic equations. For example the command

$$solve[x^\wedge 2 + (a+1)*x + a == 0, x]$$

gives the solution $x = -1$ and $x = -a$.

The replacement operator $/.$ applies rules to expressions. Consider the expression

$$x*y + x*x.$$

Then

$$x*y + x*x/.x- > 2$$

yields as output $4 + 2*y$.

Amongst others, MATHEMATICA includes the following mathematical functions: Sqrt[x] (square root, $\sqrt{x}$), exp[x] (exponential function, $e^x$), Log[x] (natural logarithm, $\ln(x)$), and the trigonometric functions Sin[x], Cos[x], Tan[x] with arguments in radians.

Predefined constants are

$$I, \ Infinity, \ Pi, \ Degree, \ Golden \ Ratio, \ E, \ Euler \ Gamma, \ Catalan$$

For other commands we refer to the user manual for MATHEMATICA (see references).

MAPLE also distinguishes between small and capital letters. The command $Sin(0.1)$ gives $sin(.1)$, whereas $sin(0.1)$ gives the desired result 0.09983341665.

In MAPLE the differentiation command is $diff()$. The input

$$diff(x^{\wedge}3 + 2 * x, x)$$

yields as output $3x^2 + 2$. The integration command is $int()$. The input

$$int(x^{\wedge}2 + 1, x)$$

yields $x^3/3 + x$.

MAPLE has two different commands for solving equations. The command

$$solve(x^{\wedge}2 + (1 + a) * x + a = 0, x)$$

solves the equation $x^2 + (1 + a)x + a = 0$ and gives the result $x = -1$ and $x = -a$. The command

$$fsolve(x^{\wedge}2 - x - 1 = 0, x)$$

solves the equation $x^2 - x - 1.0 = 0$ and gives the output $-0.618...$, $1.618...$ .

The substitution command is given by subs(). For example, the command

$$subs(x = 2, x * y + x^{\wedge}2)$$

gives $2y + 4$.

Amongst others, MAPLE includes the following mathematical functions: sqrt(x) (square root, $\sqrt{x}$), exp(x) (exponential function, $e^x$), log(x) (natural logarithm, $\ln(x)$), and the trigonometric functions sin(x), cos(x), tan(x) with arguments in radians.

Predefined constants are

$$Catalan, \quad E, \quad Pi, \quad false, \quad gamma, \quad infinity, \quad true.$$

For other commands we refer to the user manual for MAPLE (see references).

## 1. Conservation Law and Schrödinger Equation

The wave function $\psi$ satisfies the *Schrödinger equation*

$$i\hbar\frac{\partial\psi}{\partial t} = \hat{H}\psi, \tag{1}$$

where

$$\hat{H} := -\frac{\hbar^2}{2m}\Delta + V(\mathbf{r}). \tag{2}$$

In the program we show that the *conservation law*

$$\frac{\partial\rho}{\partial t} + \text{divs} = 0, \qquad \text{divs} := \frac{\partial s_1}{\partial x_1} + \frac{\partial s_2}{\partial x_2} + \frac{\partial s_3}{\partial x_3} \tag{3}$$

holds, where
$$\rho := \psi^*\psi \tag{4}$$
and

$$\mathbf{s} := \frac{\hbar}{2mi}(\psi^*\nabla\psi - \psi\nabla\psi^*). \tag{5}$$

The program is written for one space dimension, but can easily be extended to higher dimensions. In one space dimension we have

$$\Delta := \frac{\partial^2}{\partial x^2}, \tag{6}$$

$$s := \frac{\hbar}{2mi}\left(\psi^*\frac{\partial\psi}{\partial x} - \psi\frac{\partial\psi^*}{\partial x}\right). \tag{7}$$

We take into account that

$$-i\hbar\frac{\partial\psi^*}{\partial t} = \hat{H}\psi^*. \tag{8}$$

```
%conser.red;

depend psi, x, t;
depend psis, x, t;
depend rho, x, t;
depend V, x;

rho := psis*psi;      % equation (4);

r1 := df(rho,t);
r2 := sub(df(psi,t)=i*hb/(2*m)*df(psi,x,2)-i*V*psi,r1);
r3 := sub(df(psis,t)=-i*hb/(2*m)*df(psis,x,2)+i*V*psis,r2);

s := hb/(2*m*i)*(psis*df(psi,x)-psi*df(psis,x));
r4 := df(s,x);

result := r4 + r3;
```

The output is

```
r1 := df(psi,t)*psis + df(psis,t)*psi$

r2 := (df(psi,x,2)*i*psis*hb + 2*df(psis,t)*m*psi -
2*i*m*v*psi*psis)/(2*m)$

r3 := (i*hb*(df(psi,x,2)*psis - df(psis,x,2)*psi))/(2*m)$

s := (hb*(df(psi,x)*psis - df(psis,x)*psi))/(2*i*m)$

r4 := (hb*(df(psi,x,2)*psis - df(psis,x,2)*psi))/(2*i*m)$

result := 0$
```

## 2. Wave Packet and Free Schrödinger Equation

The Schrödinger equation for the free particle in one space dimension is given by

$$i\hbar \frac{\partial \psi}{\partial t} = -\frac{\hbar^2}{2m} \frac{\partial^2 \psi}{\partial x^2}. \tag{1}$$

This equation admits the solution (wave packet)

$$\psi(x,t) = \frac{B}{(1 + i\hbar t/ma^2)^{1/2}} \exp\left(-\frac{x^2 - 2ia^2 k_0 x + i\hbar k_0^2 a^2 t/m}{2a^2 \left(1 + i\hbar t/ma^2\right)}\right), \tag{2}$$

where $B$ is determined by normalizing $\psi$ for a fixed $t$, i.e.,

$$\int_{-\infty}^{\infty} \psi^*(x,t)\psi(x,t)dx = 1. \tag{3}$$

For $t = 0$ we have

$$\psi(x,0) = B \exp\left(-\frac{x^2 - 2ia^2 k_0 x}{2a^2}\right) \tag{4}$$

with the density

$$\rho(x,0) = \psi^*(x,0)\psi(x,0) = |B|^2 \exp\left(-\frac{x^2}{a^2}\right). \tag{5}$$

Thus at $t = 0$ the particle is localized with $|x| \leq a$.

In the program we show that (2) is a solution of (1). We also find $\psi^*$ and evaluate the density $\rho = \psi\psi^*$. We show that $\psi^*$ satisfies the equation

$$i\hbar \frac{\partial \psi^*}{\partial t} = \frac{\hbar^2}{2m} \frac{\partial^2 \psi^*}{\partial x^2}. \tag{6}$$

```
%wavpack.red;

depend f1, x, t;
depend f2, x, t;
depend psi, x, t;
depend rho, x, t;

f1:=B/(sqrt(1+i*hb*t/(m*a2)));
f2:=exp(-(x*x-2*i*a2*k0*x+i*hb*t*k0*k0*a2/m)/(2*a2*(1+i*hb*t/(m*a2))));
psi := f1*f2;
res1 := hb*hb*df(psi,x,2)/(2*m);
res2 := i*hb*df(psi,t);
result1 := res1 + res2;

depend f1s, x, t;
depend f2s, x, t;
depend psis, x, t;

f1s:=sub(i=-i,f1);
f2s:=sub(i=-i,f2);
psis := f1s*f2s;
res3 := hb*hb*df(psis,x,2)/(2*m);
res4 := -i*hb*df(psis,t);
result2 := res3 + res4;

rho := psis*psi;

result3 := df(rho,t)+hb/(2*m*i)*df((psis*df(psi,x)-psi*df(psis,x)),x);
```

The output is obviously

$$\text{result1} = 0, \qquad \text{result2} = 0, \qquad \text{result3} = 0.$$

## 3. Separation Ansatz and Schrödinger Equation

The Schrödinger equation is given by

$$i\hbar\frac{\partial\psi}{\partial t} = \hat{H}\psi,$$  (1)

where $\hat{H}$ is the Hamilton operator. We consider the case

$$\hat{H}\psi := -\frac{\hbar^2}{2m}\frac{\partial^2\psi}{\partial x^2} + V(x)\psi.$$  (2)

We make the so-called *separation ansatz* (or *product ansatz*)

$$\psi(x,t) = T(t)u(x).$$  (3)

Inserting this ansatz into the Schrödinger equation yields

$$i\hbar u(x)\frac{dT}{dt} = -\frac{\hbar^2}{2m}T(t)\frac{d^2u}{dx^2} + V(x)T(t)u(x).$$  (4)

It follows that

$$i\hbar\frac{1}{T}\frac{dT}{dt} = -\frac{\hbar^2}{2m}\frac{1}{u(x)}\frac{d^2u}{dx^2} + V(x),$$  (5)

since $\hat{H}$ does not depend on $t$. This can only be satisfied if both sides are equal to a constant (dimension : energy). Consequently

$$i\hbar\frac{dT}{dt} = ET(t)$$  (6)

and

$$-\frac{\hbar^2}{2m}\frac{d^2u}{dx^2} + V(x)u(x) = Eu(x).$$  (7)

Equation (7) is called the eigenvalue equation of the Schrödinger equation.

In the program we insert the separation ansatz (3) into (1) and separate out the space and time dependent terms. Equation (7) is the eigenvalue equation. The solution of (6) is given by

$$T(t) = C \exp(-iEt/\hbar). \tag{8}$$

```
%separ.red;

depend psi, x, t;
depend u, x;
depend Tm, t;
depend V, x;

psi := Tm*u;    %Product ansatz;

% Schroedinger equation;
r1 := i*hb*df(psi,t) + df(psi,x,2)*(hb**2)/(2*m) - V*psi;

on div;
r2 := r1/(Tm*u);

r3 := coeffn(r2,hb,2) + coeffn(r2,hb,0);
r4 := coeffn(r2,hb,1);
```

The output is

```
r1:=(df(u,x,2)*tm*hb**2+2*df(tm,t)*i*m*u*hb-2*m*u*v*tm)/(2*m)$

r2:=1/2*df(u,x,2)*m**(-1)*u**(-1)*hb**2+df(tm,t)*i*tm**(-1)*hb-v$

r3 := 1/2*df(u,x,2)*m**(-1)*u**(-1) - v$

r4 := df(tm,t)*i*tm**(-1)$
```

## 4. Matrix Representation in the Hilbert Space $L_2(-\pi, \pi)$

Consider the Hilbert space $L_2(-\pi, \pi)$ and the linear bounded operator $A$

$$Af := xf(x), \qquad f \in L_2(-\pi, \pi) \tag{1}$$

in this Hilbert space. In the Hilbert space $L_2(-\pi, \pi)$ a basis $\mathcal{B}$ is given by

$$\mathcal{B} := \left\{ \phi_k(x) := \frac{1}{\sqrt{2\pi}} \exp(ikx), \quad k \in \mathcal{Z} \right\}. \tag{2}$$

We evaluate the matrix representation of the operator $A$. For $k = l$ we have

$$A_{kk} = \langle \phi_k, x\phi_k \rangle = \frac{1}{2\pi} \int_{-\pi}^{\pi} e^{ikx} x e^{-ikx} dx = \frac{1}{2\pi} \int_{-\pi}^{\pi} x\,dx = 0. \tag{3}$$

For $k \neq l$ we find that

$$A_{kl} = \frac{1}{2\pi} \int_{-\pi}^{\pi} e^{ikx} x e^{-ilx} dx = \frac{i}{(l-k)\cos((k+l)\pi)}. \tag{4}$$

Introducing the order $0, 1, -1, 2, -2, \ldots$ we find that the infinite dimensional matrix $A$ has the form

$$A = \begin{pmatrix} 0 & -i & i & i/2 & \cdots \\ i & 0 & -i/2 & i/2 & \cdots \\ -i & i/2 & 0 & -i & \cdots \\ -i/2 & -i/2 & \ddots & & \ddots \\ & & & \ddots & & \ddots \\ & & \ddots & & & \ddots \end{pmatrix}. \tag{5}$$

In the program we evaluate the matrix representation of $A$. We consider the cases $k = l$ and $k \neq l$ separately.

```
%matrep.red;
depend F, X;
depend G, X;
F := exp(i*K*X)/sqrt(2*PI);
G := exp(-i*L*X)/sqrt(2*PI);
%case : K = L;
L := K;
R1 := int(F*X*G,X);
R2 := sub(x=pi,r1)-sub(X=-PI,R1);
%case : k \neq l;
clear k, l, r1, r2;
R1 := int(F*X*G,X);
R2 := sub(X=PI,R1)-sub(X=-PI,R1);
LET E**(2*I*K*PI)=1;   R2;
LET E**(2*I*L*PI)=1;   R2;
LET E**(I*K*PI+I*L*PI) = COS(PI*(L+K));   R2;
%special case : L = K + 1;
L := K + 1;
R2;
%special case : K = L + 1;
clear K, L, R2;
R1 := int(F*X*G,X);
R2 := sub(X=PI,R1)-sub(X=-PI,R1);
K := L + 1;
R2;
LET COS(2*L*PI+PI) = -1;
R2;
```

The output is

```
%case : K = L;
R2 := 0$
%case : k \neq l;
R2 := ( - E**(2*I*K*PI)*I*K*PI + E**(2*I*K*PI)*I*L*PI + E**(2
*I*K*PI) - E**(2*I*L*PI)*I*K*PI + E**(2*I*L*PI)*I*L*PI -
E**(2*I*L*PI))/(2*E**(I*K*PI + I*L*PI)*PI*(K**2 - 2*K*L + L**2))$

( - I)/(COS(K*PI + L*PI)*(K - L))$

%special case : L = K + 1;
-I$
%special case : K = L + 1;
I$
```

## 5. One-Dimensional Potential and Trial Function

The eigenvalue equation in one space dimension is given by

$$-\frac{\hbar^2}{2m}\frac{d^2u}{dx^2} + V(x)u(x) = Eu(x). \tag{1}$$

We use the variational principle to estimate the ground state energy of a particle in the potential

$$V(x) = \begin{cases} \infty & \text{for } x < 0 \\ cx & \text{for } x > 0, \end{cases} \tag{2}$$

where $c > 0$. Owing to this potential the spectrum is discrete and bounded from below. We use

$$u(x) = \begin{cases} x\exp(-ax) & \text{for } x > 0 \\ 0 & \text{for } x < 0 \end{cases} \tag{3}$$

as a trial function, where $a > 0$. We have to keep in mind that the trial function (3) is not yet normalized. From (1) we find that the expectation value for the energy is given by

$$\langle E \rangle := \frac{\langle u|\hat{H}|u\rangle}{\langle u|u\rangle} = \frac{\int_0^\infty xe^{-ax}\left(-\frac{\hbar^2}{2m}\frac{d^2}{dx^2} + cx\right)xe^{-ax}dx}{\int_0^\infty x^2\exp(-2ax)dx} = \frac{3c}{2a} + \frac{\hbar^2a^2}{2m}. \tag{4}$$

This expectation value has a minimum for

$$a = \left(\frac{3mc}{2\hbar^2}\right)^{1/3}. \tag{5}$$

In the program we evaluate (4) and then determine the minimum of $\langle E \rangle$ with respect to *a.* Thus the ground-state energy is greater than or equal to

$$\frac{9}{4}\left(\frac{2\hbar^2 c^2}{3m}\right)^{1/3}. \tag{6}$$

```
%trial.red;

depend u, x;
depend RES1, x;

u := x*exp(-a*x);
RES1 := -(hb)*(hb)/(2*m)*df(u,x,2) + c*x*u;
RES2 := int(u*RES1,x);
RES3 := -sub(x=0,RES2);
RES4 := int(u*u,x);
norm := -sub(x=0,RES4);

EXPE := RES3/norm;
MINIM := df(EXPE,A);
RES5 := SOLVE(MINIM=0,A);
```

The output is

```
RES1:=(-A**2*X*HB**2+2*A*HB**2+2*C*M*X**2)/(2*E**(A*X)*M)$
RES2:=(2*A**5*X**2*HB**2-2*A**4*X*HB**2-4*A**3*C*M*X**3
-A**3*HB**2-6*A**2*C*M*X**2-6*A*C*M*X-3*C*M)/(8*E**(2*A*X)*A**4*M)$
RES3 := (A**3*HB**2 + 3*C*M)/(8*A**4*M)$
RES4 := ( - 2*A**2*X**2 - 2*A*X - 1)/(4*E**(2*A*X)*A**3)$
NORM := 1/(4*A**3)$
EXPE := (A**3*HB**2 + 3*C*M)/(2*A*M)$
MINIM := (2*A**3*HB**2 - 3*C*M)/(2*A**2*M)$
RES5 := {A=( - (M**(1/3)*C**(1/3)*3**(1/3))*(SQRT(3)*I + 1))/
(2*HB**(2/3)*2**(1/3)),
A=(M**(1/3)*C**(1/3)*3**(1/3)*(SQRT(3)*I - 1))/(2*HB
**(2/3)*2**(1/3)),
A=(M**(1/3)*C**(1/3)*3**(1/3))/(HB**(2/3)*2**(1/3))}$
```

*Remark:* Only the real solution of RES5 is valid in our case.

## 6. Heisenberg Equation of Motion

The *Heisenberg equation of motion* is given by

$$i\hbar\frac{d\hat{A}}{dt} = [\hat{A}, \hat{H}](t),\tag{1}$$

where $[\,,\,]$ denotes the *commutator*, i.e. $[X,Y] := XY - YX$. Let

$$\hat{H} = \hbar\omega\sigma_z,\tag{2}$$

where

$$\sigma_z := \begin{pmatrix} 1 & 0 \\ 0 & -1 \end{pmatrix}.\tag{3}$$

*Remark:* The matrices $\sigma_x$, $\sigma_y$ and $\sigma_z$ are the *Pauli matrices*, where

$$\sigma_x := \begin{pmatrix} 0 & 1 \\ 1 & 0 \end{pmatrix}, \qquad \sigma_y := \begin{pmatrix} 0 & -i \\ i & 0 \end{pmatrix}.\tag{4}$$

We evaluate the time evolution of $\sigma_x$. Since

$$[\sigma_x, \hat{H}] = \hbar\omega[\sigma_x, \sigma_z] = -2i\hbar\omega\sigma_y\tag{5}$$

we obtain

$$\frac{d\sigma_x}{dt} = -2\omega\sigma_y(t).\tag{6}$$

Now we have to evaluate the time evolution of $\sigma_y$, i.e.,

$$i\hbar\frac{d\sigma_y}{dt} = [\sigma_y, \hat{H}](t).\tag{7}$$

Since

$$[\sigma_y, \hat{H}] = \hbar\omega[\sigma_y, \sigma_z] = 2i\hbar\omega\sigma_x\tag{8}$$

we find

$$\frac{d\sigma_y}{dt} = 2\omega\sigma_x(t). \tag{9}$$

To summarize:

$$\frac{d\sigma_x}{dt} = -2\omega\sigma_y(t). \tag{10a}$$

$$\frac{d\sigma_y}{dt} = 2\omega\sigma_x(t). \tag{10b}$$

The initial conditions are

$$\sigma_x(t = 0) = \sigma_x, \qquad \sigma_y(t = 0) = \sigma_y. \tag{11}$$

Then the solution of the initial value problem is given by

$$\sigma_x(t) = \sigma_x \cos(2\omega t) - \sigma_y \sin(2\omega t) \tag{12a}$$

$$\sigma_y(t) = \sigma_y \cos(2\omega t) + \sigma_x \sin(2\omega t). \tag{12b}$$

*Remark:* The solution of the Heisenberg equation of motion can also be given by

$$\sigma_x(t) = \exp(i\hat{H}t/\hbar)\sigma_x \exp(-i\hat{H}t/\hbar) \tag{13a}$$

$$\sigma_y(t) = \exp(i\hat{H}t/\hbar)\sigma_y \exp(-i\hat{H}t/\hbar). \tag{13b}$$

Since

$$\exp(i\hat{H}t/\hbar) := \sum_{k=0}^{\infty} \frac{1}{k!}\left(\frac{i\hat{H}t}{\hbar}\right)^k = \sum_{k=0}^{\infty} \frac{1}{k!}(i\omega\sigma_z)^k \tag{14}$$

and

$$\sigma_z^2 = I_2, \tag{15}$$

where $I_2$ is the $2 \times 2$ unit matrix, we find that

$$\exp(i\hat{H}t/\hbar) = \begin{pmatrix} \exp(i\omega t) & 0 \\ 0 & \exp(-i\omega t) \end{pmatrix} \tag{16a}$$

and

$$\exp(-i\hat{H}t/\hbar) = \begin{pmatrix} \exp(-i\omega t) & 0 \\ 0 & \exp(i\omega t) \end{pmatrix}.$$ (16b)

*Remark:* The Heisenberg equation of motion

$$i\hbar\frac{d\sigma_x}{dt} = [\sigma_x, \hat{H}](t) = \hbar\omega[\sigma_x, \sigma_z](t)$$ (17)

can be brought into a dimensionless form when we set

$$\tau(t) = \omega t, \qquad \tilde{\sigma}_x(\tau(t)) = \sigma_x(t).$$ (18)

Then we have

$$i\frac{d\tilde{\sigma}_x}{d\tau} = [\tilde{\sigma}_x, \tilde{\sigma}_z](\tau)$$ (19a)

and

$$i\frac{d\tilde{\sigma}_y}{d\tau} = [\tilde{\sigma}_y, \tilde{\sigma}_z](\tau).$$ (19b)

In the program we calculate the time evolution of $\sigma_x$ and $d\sigma_x/dt$, where $d\sigma_x/dt = -2\omega\sigma_y(t)$. We use the following notation:

$$\frac{d\sigma_x}{dt} \rightarrow sxt, \qquad \frac{d^2\sigma_x}{dt^2} \rightarrow sxtt.$$ (20)

```
%heisen.red;

matrix H(2,2);
matrix sx(2,2);
matrix sy(2,2);
matrix sz(2,2);

sx(1,1) := 0; sx(1,2) := 1; sx(2,1) := 1; sx(2,2) := 0;
sy(1,1) := 0; sy(1,2) := -i; sy(2,1) := i; sy(2,2) := 0;
sz(1,1) := 1; sz(1,2) := 0; sz(2,1) := 0; sz(2,2) := -1;

H := hb*om*sz;

sxt := 1/(i*hb)*(sx*H - H*sx);
sxtt := 1/(i*hb)*(sxt*H - H*sxt);
```

The output is given by

```
H := MAT((HB*OM,0),(0, - HB*OM))$

SXT := MAT((0,( - 2*OM)/I),((2*OM)/I,0))$
SXTT := MAT((0, - 4*OM**2),( - 4*OM**2,0))$
```

## 7. Harmonic Oscillator

The eigenvalue equation for the one-dimensional harmonic oscillator is given by $\hat{H}u = Eu$, where

$$\hat{H} = -\frac{\hbar^2}{2m}\frac{d^2}{dx^2} + \frac{1}{2}m\omega^2 x^2 \tag{1}$$

and $u \in L_2(\mathcal{R})$. The eigenvalue equation can be written as

$$\frac{\hbar}{\omega m}\frac{d^2 u}{dx^2} - \frac{m\omega x^2}{\hbar}u + \frac{2E}{\hbar\omega}u = 0. \tag{2}$$

Introducing the quantities ($x_0$ has the dimension of a length)

$$x_0 := \sqrt{\frac{\hbar}{m\omega}}, \qquad \xi := \frac{x}{x_0} \tag{3}$$

and $u(x) = \tilde{u}(\xi(x))$ we find that

$$\frac{du}{dx} = \frac{d\tilde{u}}{d\xi}\frac{1}{x_0}, \qquad \frac{d^2 u}{dx^2} = \frac{d^2\tilde{u}}{d\xi^2}\frac{1}{x_0^2}. \tag{4}$$

We notice that the quantity $\xi$ is dimensionless. Thus we can write (2) as

$$\left(\frac{d^2}{d\xi^2} - \xi^2 + \frac{2E}{\hbar\omega}\right)\tilde{u}(\xi) = 0. \tag{5}$$

Introducing the dimensionless quantity

$$\lambda := \frac{2E}{\hbar\omega} - 1 \tag{6}$$

we arrive at

$$\left(\frac{d^2}{d\xi^2} - \xi^2 + \lambda + 1\right) \tilde{u}(\xi) = 0. \tag{7}$$

In the program we insert the ansatz

$$\tilde{u}(\xi) = H(\xi) \exp(-\xi^2/2) \tag{8}$$

into (7) and find a second-order differential equation for $H$. We obtain

$$\frac{d^2 H}{d\xi^2} - 2\xi \frac{dH}{d\xi} + \lambda H = 0. \tag{9}$$

Equation (9) is called the *Hermite differential equation.*

```
%harm1.red;

depend UT, XI;
depend H, XI;
UT := A*H*EXP(-XI**2/2);

R1 := df(UT,XI,2) - XI*XI*UT + (LAM + 1)*UT;

ON DIV;
R1 := R1*H/UT;
```

The output is

```
R1 := (A*(DF(H,XI,2) - 2*DF(H,XI)*XI + H*LAM))/E**(XI**2/2)$

R1 := DF(H,XI,2) - 2*DF(H,XI)*XI + H*LAM$
```

## 8. Harmonic Oscillator and Recursion Relation

We consider again the one-dimensional harmonic oscillator (see 7). The eigenvalue equation can be brought into the form (see 7)

$$\left(\frac{d^2}{d\xi^2} - \xi^2 + \lambda + 1\right)\tilde{u}(\xi) = 0. \tag{1}$$

Inserting the ansatz
$$\tilde{u}(\xi) = H(\xi)\exp(-\xi^2/2) \tag{2}$$

into (1) yields

$$\left(\frac{d^2}{d\xi^2} - 2\xi\frac{d}{d\xi} + \lambda\right)H(\xi) = 0. \tag{3}$$

To find the solution of (3) we consider the power series ansatz

$$H(\xi) = \sum_{m=0}^{\infty} a_m\xi^m, \tag{4}$$

where $a_m$ are the expansion coefficients.

In the program we show that the expansion coefficients $a_m$ satisfy

$$a_{m+2} = \frac{2m - \lambda}{m^2 + 3m + 2}a_m. \tag{5}$$

```
%harm2.our;

operator a;
f  := a(m)*x**(m) + a(m+1)*x**(m+1) + a(m+2)*x**(m+2);
R1 := df(f,x);
R1 := -2*R1*x;
R2 := df(f,x,2);
R2 := sub(m=m+2,R2);
R3 := R2 + R1 + lam*f;
on div;
R4 := R3/(x**m);
R5 := coeffn(R4,x,2);
R6 := sub(m=m-2,R5);
R7 := solve(R6,a(m+2));
```

The output is given by

```
R4 := A(M + 4)*M**2*X**2 + 7*A(M + 4)*M*X**2 + 12*A(M + 4)*X**2 +
A(M + 3)*M**2*X + 5*A(M + 3)*M*X + 6*A(M + 3)*X + A(M + 2)*M**2 -
2*A(M + 2)*M*X**2 + 3*A(M + 2)*M + A(M + 2)*X**2*LAM -
4*A(M + 2)*X**2 + 2*A(M + 2) - 2*A(M + 1)*M*X + A(M + 1)*X*LAM -
2*A(M + 1)*X - 2*A(M)*M + A(M)*LAM$

R5 := A(M + 4)*M**2 + 7*A(M + 4)*M + 12*A(M + 4) - 2*A(M + 2)*M +
A(M + 2)*LAM - 4*A(M + 2)$

R6 := A(M + 2)*M**2 + 3*A(M + 2)*M + 2*A(M + 2) - 2*A(M)*M + A(M)*LAM$

R7 := {A(M + 2)=(A(M)*(2*M - LAM))/(M**2 + 3*M + 2)}$
```

## 9. Commutation Relations of $\hat{\mathbf{p}}$ and $\hat{\mathbf{q}}$

In nonrelativistic quantum mechanics, one usual representation for a system of $N$ particles moving in a potential $V$ is

$$\mathcal{H} = L_2(\mathcal{R}^{3N}). \tag{1}$$

This choice is called the Schrödinger representation (as distinct from the Schrödinger picture). The function $\psi(\mathbf{q}) \in \mathcal{H}$ has the interpretation of giving the probability distribution

$$\rho(\mathbf{q}) := |\psi(\mathbf{q})|^2 \tag{2}$$

for the position of the particle in $\mathcal{R}^{3N}$. Using the postulate that the momentum operator $\hat{\mathbf{p}}$ is the infinitesimal generator of the unitary space translation group, we find

$$p_{kj} \rightarrow \hat{p}_{kj} = -i\hbar \frac{\partial}{\partial q_{kj}} \tag{3}$$

and a nonrelativistic Hamilton function of the form

$$H = \sum_{k=1}^{N} \sum_{j=1}^{3} \frac{p_{kj}^2}{2m_k} + V(\mathbf{q}) \tag{4}$$

becomes the elliptic differential operator (Hamilton operator)

$$\hat{H} = -\sum_{k=1}^{N} \sum_{j=1}^{3} \frac{\hbar^2}{2m_k} \frac{\partial^2}{\partial q_{kj}^2} + V(\mathbf{q}). \tag{5}$$

In other words the Hamilton operator $\hat{H}$ follows from $H$ via the *quantization*

$$p_{kj} \rightarrow -i\hbar \frac{\partial}{\partial q_{kj}}, \qquad q_{kj} \rightarrow \hat{q}_{kj}. \tag{6}$$

The operator $\hat{q}_{kj}$ is defined by $\hat{q}_{kj} f(\mathbf{q}) := q_{kj} f(\mathbf{q})$. We find the (canonical) *commutation relations*

$$[\hat{q}_{kj}, \hat{q}_{k'j'}] = 0, \qquad [\hat{p}_{kj}, \hat{p}_{k'j'}] = 0, \qquad [\hat{p}_{kj}, \hat{q}_{k'j'}] = -i\hbar \delta_{kk'} \delta_{jj'} I. \tag{7}$$

They are preserved by the Heisenberg equation of motion. The unit of action $\hbar$ has the same dimension as **pq**.

```
%momen.our;

operator x, p, comm, delta$
noncom x, p$
antisymmetric comm$

%Definition of Kronecker delta;
let { delta(~a,~b) => 0
      when numberp a and numberp b and a neq b,
      delta(~a,~a) => 1 }$

%Definition of commutation relation;
let { comm(x(~a),x(~b)) => 0,
      comm(p(~a),p(~b)) => 0,
      comm(x(~a),p(~b)) => hb*i*delta(a,b),
      comm(~a+~b,~c) => comm(a,c) + comm(b,c),
      comm(~a**2,~b) => a*comm(a,b) + comm(a,b)*a }$

x2 := for k:=1:3 sum x(k)**2;
p2 := for k:=1:3 sum p(k)**2;

for k := 1:3 collect comm(x2,p(k));
for k := 1:3 collect comm(p2,x(k));
```

The output is given by

```
x2 := x(3)**2 + x(2)**2 + x(1)**2$
p2 := p(3)**2 + p(2)**2 + p(1)**2$

for k := 1:3 collect comm(x2,p(k));
{2*i*hb*x(1),2*i*hb*x(2),2*i*hb*x(3)}$

for k := 1:3 collect comm(p2,x(k));
{ - 2*i*hb*p(1), - 2*i*hb*p(2), - 2*i*hb*p(3)}$
```

## 10. Anharmonic Oscillator

The eigenvalue equation in one space dimension is given by

$$-\frac{d^2u}{dx^2} + V(x)u(x) = Eu(x), \tag{1}$$

where

$$V \to \frac{2mV}{\hbar^2}, \qquad E \to \frac{2mE}{\hbar^2}. \tag{2}$$

The underlying Hilbert space is $L_2(\mathcal{R})$. Let

$$V(x) = a^2x^6 - 3x^2, \qquad a > 0. \tag{3}$$

We show that

$$u(x) = \exp(-\frac{1}{4}ax^4) \tag{4}$$

is an exact solution of (1) with the potential given by (3). We also find the energy eigenvalue

$$E_0 = 0. \tag{5}$$

Obviously, the wave function (4) is the ground state, because it has no nodes.

We also consider the generalized case

$$V(x) = a^2x^6 + 2abx^4 + (b^2 - 3a)x^2 \tag{6}$$

with the exact eigenfunction

$$u(x) = \exp(-ax^4/4 - bx^2/2). \tag{7}$$

The eigenvalue is

$$E(a,b) = b. \tag{8}$$

```
%analy.red;

depend u1, x;
depend V1, x;

u1 := exp(-(a*x**4)/4);
V1 := a*a*x**6 - 3*a*x*x;

res1 := -df(u1,x,2) + V1*u1 - En*u1;
res2 := res1/u1;
res3 := solve(res2,En);

%generalized case;
depend u2, x;
depend V2, x;

u2 := exp(-(a*x**4)/4 - (b*x**2)/2);
V2 := a*a*x**6 + 2*a*b*x**4 + (b*b - 3*a)*x**2;

res4 := -df(u2,x,2) + V2*u2 - En*u2;
res5 := res4/u2;
res6 := solve(res5,En);
```

The output is given by

```
RES1 := ( - EN)/E**((A*X**4)/4)$
RES2 :=  - EN$
RES3 := {EN=0}$

%generalized case;
RES4 := (B - EN)/E**((A*X**4 + 2*B*X**2)/4)$
RES5 := B - EN$
RES6 := {EN=B}$
```

## 11. One-Dimensional WKB-Solutions

The one-dimensional eigenvalue equation can be written in the form

$$\frac{d^2u}{dx^2} + \frac{2m}{\hbar^2}(E - V(x))u = 0. \tag{1}$$

We set

$$u(x) := \exp(iw/\hbar), \qquad w(x) := S(x) + \frac{\hbar}{i}\ln A(x). \tag{2}$$

We note that $S$ and $\ln A$ are even functions of $\hbar$. Inserting (2) into (1) we obtain the system of equations

$$\left(\frac{dS}{dx}\right)^2 - 2m(E - V) = \hbar^2 \frac{1}{A}\frac{d^2A}{dx^2}, \qquad 2\frac{dA}{dx}\frac{dS}{dx} + A\frac{d^2S}{dx^2} = 0. \tag{3}$$

The second equation of (3) (equation of continuity) can be integrated to yield

$$A(x) = C\left(\frac{dS}{dx}\right)^{-1/2}. \tag{4}$$

Substituting this expression for $A$ into the first equation of (3), we obtain the equation

$$\left(\frac{dS}{dx}\right)^2 = 2m(E - V) + \hbar^2\left(\frac{3}{4}\left(\frac{\frac{d^2S}{dx^2}}{\frac{dS}{dx}}\right)^2 - \frac{1}{2}\frac{\frac{d^3S}{dx^3}}{\frac{dS}{dx}}\right). \tag{5}$$

This nonlinear differential equation of third order in $S$ is equivalent to the initial eigenvalue equation.

In the WKB approximation we expand $S$ in a power series in $\hbar^2$

$$S = S_0 + \hbar^2 S_1 + \hbar^4 S_2 + \dots \tag{6}$$

In the program we insert (2) into (1) to evaluate system (3).

```
%wkb.red

depend u, x;
depend S, x;
depend A, x;
depend V, x;

u := exp(i*(S-i*hb*log(A))/hb);     % equation 2;

% Eigenvalue equation 1;
R1 := df(u,x,2) + (2*m)*(EN - V)*u;

R2 := coeffn(R1,i,1);
R3 := coeffn(R2,exp(i*S/hb),1);
R4 := coeffn(R1,i,0);
R5 := coeffn(R4,exp(i*S/hb),1);
```

The output is

```
r1 := (e**((i*s)/hb)*(df(a,x,2)*hb**2 + 2*df(a,x)*df(s,x)*i*hb +
df(s,x,2)*a*i*hb - df(s,x)**2*a - 2*a*m*v*hb**2 +
2*a*m*hb**2*en))/hb**2$

r2 := (e**((i*s)/hb)*(2*df(a,x)*df(s,x) + df(s,x,2)*a))/hb$

r3 := (2*df(a,x)*df(s,x) + df(s,x,2)*a)/hb$

r4 := (e**((i*s)/hb)*(df(a,x,2)*hb**2 - df(s,x)**2*a - 2*a*m*v*hb**2 +
2*a*m*hb**2*en))/hb**2$

r5 := (df(a,x,2)*hb**2 - df(s,x)**2*a - 2*a*m*v*hb**2 +
2*a*m*hb**2*en)/hb**2$
```

## 12. Angular Momentum Operators I

In the classical case the *angular momentum* is given by

$$\mathbf{L} := \mathbf{r} \times \mathbf{p}, \tag{1}$$

where $\times$ denotes the cross product. The components of $\mathbf{L}$ are given by

$$L_x := yp_z - zp_y, \qquad L_y := zp_x - xp_z, \qquad L_z := xp_y - yp_x. \tag{2}$$

Introducing the quantization

$$p_x \rightarrow -i\hbar \frac{\partial}{\partial x}, \qquad p_y \rightarrow -i\hbar \frac{\partial}{\partial y}, \qquad p_z \rightarrow -i\hbar \frac{\partial}{\partial z} \tag{3}$$

yields

$$\hat{L}_x := \frac{\hbar}{i}\left(y\frac{\partial}{\partial z} - z\frac{\partial}{\partial y}\right), \quad \hat{L}_y := \frac{\hbar}{i}\left(z\frac{\partial}{\partial x} - x\frac{\partial}{\partial z}\right), \quad \hat{L}_z := \frac{\hbar}{i}\left(x\frac{\partial}{\partial y} - y\frac{\partial}{\partial x}\right). \tag{4}$$

The angular momentum operators $\hat{L}_x$, $\hat{L}_y$ and $\hat{L}_z$ form a basis of a Lie algebra under the commutator. The commutators are given by

$$[\hat{L}_x, \hat{L}_y] = i\hbar\hat{L}_z, \qquad [\hat{L}_z, \hat{L}_x] = i\hbar\hat{L}_y, \qquad [\hat{L}_y, \hat{L}_z] = i\hbar\hat{L}_x. \tag{5}$$

where $[\,,\,]$ denotes the commutator.

In the program we evaluate the commutators. We consider the angular momentum operators as vector fields. We apply the notation $x = x(1)$, $y = x(2)$ and $z = x(3)$.

```
%lxlylz.red;

operator LX, LY, LZ, R12, x;
depend LX(j), x(1), x(2), x(3);
depend LY(j), x(1), x(2), x(3);
depend LZ(j), x(1), x(2), x(3);
depend R12(j), x(1), x(2), x(3);
LX(1) := 0;
LX(2) := i*hb*x(3);
LX(3) := -i*hb*x(2);
LY(1) := -i*hb*x(3);
LY(2) := 0;
LY(3) := i*hb*x(1);
LZ(1) := i*hb*x(2);
LZ(2) := -i*hb*x(1);
LZ(3) := 0;

%Commutator of LX and LY;
for k := 1:3 do
R12(k) := for j := 1:3 sum
LX(j)*df(LY(k),x(j)) - LY(j)*df(LX(k),x(j));

for k := 1:3 do
write R12(k);
```

The output is

```
- X(2)*HB**2$
X(1)*HB**2$
0$
```

## 13. Angular Momentum Operators II

The angular momentum operators in Cartesian coordinates are given by

$$\hat{L}_x := \frac{\hbar}{i}\left(y\frac{\partial}{\partial z} - z\frac{\partial}{\partial y}\right), \quad \hat{L}_y := \frac{\hbar}{i}\left(z\frac{\partial}{\partial x} - x\frac{\partial}{\partial z}\right), \quad \hat{L}_z := \frac{\hbar}{i}\left(x\frac{\partial}{\partial y} - y\frac{\partial}{\partial x}\right). \quad (1)$$

The commutators are given by

$$[\hat{L}_x, \hat{L}_y] = i\hbar\hat{L}_z, \qquad [\hat{L}_z, \hat{L}_x] = i\hbar\hat{L}_y, \qquad [\hat{L}_y, \hat{L}_z] = i\hbar\hat{L}_x. \qquad (2)$$

From $\hat{L}_x$, $\hat{L}_y$ and $\hat{L}_z$ we can construct the so-called *Casimir operator*

$$\hat{L}^2 := \hat{L}_x^2 + \hat{L}_y^2 + \hat{L}_z^2. \qquad (3)$$

We have

$$[\hat{L}^2, \hat{L}_x] = 0, \qquad [\hat{L}^2, \hat{L}_y] = 0, \qquad [\hat{L}^2, \hat{L}_z] = 0. \qquad (4)$$

In the program we show that

$$[\hat{L}_x, \hat{L}_y] = i\hbar\hat{L}_z. \qquad (5)$$

Furthermore, we determine $\hat{L}^2$ and show that

$$[\hat{L}^2, \hat{L}_x] = 0. \qquad (6)$$

We consider the operators $\hat{L}_x$, $\hat{L}_y$, $\hat{L}_z$ as differential operators. This approach is more general than the approach used in the previous program. We apply the notation $x = x(1)$, $y = x(2)$ and $z = x(3)$.

```
%loper.red;

operator LX, LY, LZ, u, x, A, B, C, L2, res;
depend u, x(1), x(2), x(3);
LX := -i*hb*(x(2)*df(u,x(3)) - x(3)*df(u,x(2)))$
LY := -i*hb*(x(3)*df(u,x(1)) - x(1)*df(u,x(3)))$
LZ := -i*hb*(x(1)*df(u,x(2)) - x(2)*df(u,x(1)))$

%Commutator of LX and LY$
A := sub(u=LY,LX)$
B := sub(u=LX,LY)$
write "The commutator is: ";
C := A - B;

%Evaluation of L2:=LX**2+LY**2+LZ**2$
L2 := sub(u=LX,LX) + sub(u=LY,LY) + sub(u=LZ,LZ);

%Commutator of L2 and LX$
res := sub(u=LX,L2) - sub(u=L2,LX);
```

The output is

```
C := HB**2*( - X(2)*DF(U,X(1)) + X(1)*DF(U,X(2)))$

L2 := HB**2*( - X(3)**2*DF(U,X(2),2) - X(3)**2*DF(U,X(1),2) +
2*X(3)*X(2)*DF(U,X(3),X(2)) + 2*X(3)*X(1)*DF(U,X(3),X(1)) +
2*X(3)*DF(U,X(3)) - X(2)**2*DF(U,X(3),2) - X(2)**2*DF(U,X(1),2) +
2*X(2)*X(1)*DF(U,X(2),X(1)) + 2*X(2)*DF(U,X(2)) -
X(1)**2*DF(U,X(3),2) - X(1)**2*DF(U,X(2),2) + 2*X(1)*DF(U,X(1)))$

RES := 0$
```

## 14. Angular Momentum Operators III

The operators $\hat{L}_+$ and $\hat{L}_-$ are defined as

$$\hat{L}_+ := \hat{L}_x + i\hat{L}_y, \qquad \hat{L}_- := \hat{L}_x - i\hat{L}_y. \tag{1}$$

The *spherical coordinates* are given by

$$x = r\sin\theta\cos\phi, \qquad y = r\sin\theta\sin\phi, \qquad z = r\cos\theta, \tag{2}$$

where $0 \le \phi < 2\pi$ and $0 \le \theta \le \pi$. The angular momentum operators can be expressed in spherical coordinates $(\phi, \theta)$ as

$$\hat{L}_+ := \hbar \exp(i\phi)\left(\frac{\partial}{\partial\theta} + i\cot\theta\frac{\partial}{\partial\phi}\right) \tag{3a}$$

$$\hat{L}_- := \hbar \exp(-i\phi)\left(-\frac{\partial}{\partial\theta} + i\cot\theta\frac{\partial}{\partial\phi}\right) \tag{3b}$$

$$\hat{L}_z := -i\hbar\frac{\partial}{\partial\phi}. \tag{3c}$$

In the program we show that

$$[\hat{L}_+, \hat{L}_-] = 2\hbar\hat{L}_z \tag{4}$$

and

$$[\hat{L}^2, \hat{L}_+] = 0, \tag{5}$$

where

$$\hat{L}^2 = \frac{1}{2}(\hat{L}_+\hat{L}_- + \hat{L}_-\hat{L}_+) + \hat{L}_z^2 = -\hbar^2\left(\frac{\partial^2}{\partial\theta^2} + \cot\theta\frac{\partial}{\partial\theta} + \frac{1}{\sin^2\theta}\frac{\partial^2}{\partial\phi^2}\right). \tag{6}$$

```
%lplmlz.red;

operator LP, LM, LZ, u, A, B, C, L2, res;
depend u, theta, phi;
LP := hb*exp(i*phi)*(df(u,theta)+i*(cos(theta)/sin(theta))*df(u,phi))$
LM := hb*exp(-i*phi)*(-df(u,theta)+i*(cos(theta)/sin(theta))*df(u,phi))$
LZ := -i*hb*df(u,phi)$

for all q let (cos(q))**2 = 1 - (sin(q))**2$

%Commutator of LM and LP$
A := sub(u=LP,LM)$
B := sub(u=LM,LP)$
write "The commutator is: ";
C := A - B;

%Evaluation of L2:=LX**2+LY**2+LZ**2$
L2 := (sub(u=LP,LM))/2 + (sub(u=LM,LP))/2 + sub(u=LZ,LZ);

%Commutator of L2 and LP$
res := sub(u=LP,L2) - sub(u=L2,LP);
```

The output is

```
C := 2*DF(U,PHI)*I*HB**2$
L2 := ( - HB**2*(DF(U,THETA,2)*SIN(THETA)**2 + DF(U,THETA)*
SIN(THETA)*COS(THETA) + DF(U,PHI,2)))/SIN(THETA)**2$
%Commutator of L2 and LP$
RES := 0$
```

## 15. Lie Algebra su(3) and Commutation Relations

The Lie algebra $su(3)$ of the Lie group $SU(3)$ has eight elements $L_j$ $(j = 1, 2, \ldots, 8)$, satisfying

$$[L_j, L_k] = i \sum_{n=1}^{8} c_{jkn} L_n, \tag{1}$$

where $c_{jkn}$ are real constants completely antisymmetric in $j$, $k$, $n$. The basic represention is 3, in which the generators are written in the form

$$L_j = \frac{1}{2}\lambda_j, \qquad j = 1, 2, \ldots, 8 \tag{2}$$

where $\lambda_j$ are $3 \times 3$ matrices. They act on basis vectors of the form

$$\mathbf{x} = \begin{pmatrix} x_1 \\ x_2 \\ x_3 \end{pmatrix}. \tag{3}$$

An infinitesimal element of the group is represented by the transformation

$$\mathbf{x}' = S\mathbf{x}, \tag{4}$$

where

$$S = I - \frac{i}{2} \sum_{j=1}^{8} \epsilon_j \lambda_j. \tag{5}$$

Here $\epsilon_j$ $(j = 1, 2, \ldots, 8)$ are arbitrary infinitesimal real numbers. A set of hermitian matrices satisfying (1) is given by

$$\lambda_1 = \begin{pmatrix} 0 & 1 & 0 \\ 1 & 0 & 0 \\ 0 & 0 & 0 \end{pmatrix}, \qquad \lambda_2 = \begin{pmatrix} 0 & -i & 0 \\ i & 0 & 0 \\ 0 & 0 & 0 \end{pmatrix}$$

$$\lambda_3 = \begin{pmatrix} 1 & 0 & 0 \\ 0 & -1 & 0 \\ 0 & 0 & 0 \end{pmatrix}, \qquad \lambda_4 = \begin{pmatrix} 0 & 0 & 1 \\ 0 & 0 & 0 \\ 1 & 0 & 0 \end{pmatrix}$$

$$\lambda_5 = \begin{pmatrix} 0 & 0 & -i \\ 0 & 0 & 0 \\ i & 0 & 0 \end{pmatrix}, \qquad \lambda_6 = \begin{pmatrix} 0 & 0 & 0 \\ 0 & 0 & 1 \\ 0 & 1 & 0 \end{pmatrix}$$

$$\lambda_7 = \begin{pmatrix} 0 & 0 & 0 \\ 0 & 0 & -i \\ 0 & i & 0 \end{pmatrix}, \qquad \lambda_8 = \frac{1}{\sqrt{3}} \begin{pmatrix} 1 & 0 & 0 \\ 0 & 1 & 0 \\ 0 & 0 & -2 \end{pmatrix}. \qquad (6)$$

Notice that $\text{tr}(\lambda_j) = 0$ for $j = 1, 2, \ldots, 8$.

In the program we implement the matrices $\lambda_1$, $\lambda_2$, ..., $\lambda_8$ and evaluate the commutator of $\lambda_1$ and $\lambda_2$.

```
%su3.red;

matrix L1(3,3), L2(3,3), R1(3,3);

L1 := mat((0,1,0),(1,0,0),(0,0,0));
L2 := mat((0,-i,0),(i,0,0),(0,0,0));
L3 := mat((1,0,0),(0,-1,0),(0,0,0));
L4 := mat((0,0,1),(0,0,0),(1,0,0));
L5 := mat((0,0,-i),(0,0,0),(i,0,0));
L6 := mat((0,0,0),(0,0,1),(0,1,0));
L7 := mat((0,0,0),(0,0,-i),(0,i,0));
L8 := mat((1,0,0),(0,1,0),(0,0,-2))/sqrt(3);

R1 := L1*L2 - L2*L1;
```

The output is

```
R1 := MAT((2*I,0,0),(0, - 2*I,0),(0,0,0))$
```

## 16. Spin-1 Lie Algebra and Commutation Relations

We introduce the matrices to describe a particle with spin-1

$$
s_x := \frac{\hbar}{\sqrt{2}} \begin{pmatrix} 0 & 1 & 0 \\ 1 & 0 & 1 \\ 0 & 1 & 0 \end{pmatrix}, \qquad s_y := \frac{\hbar}{\sqrt{2}} \begin{pmatrix} 0 & -i & 0 \\ i & 0 & -i \\ 0 & i & 0 \end{pmatrix}, \qquad s_z := \hbar \begin{pmatrix} 1 & 0 & 0 \\ 0 & 0 & 0 \\ 0 & 0 & -1 \end{pmatrix} \tag{1}
$$

$$
s_+ = \sqrt{2}\hbar \begin{pmatrix} 0 & 1 & 0 \\ 0 & 0 & 1 \\ 0 & 0 & 0 \end{pmatrix}, \qquad s_- = \sqrt{2}\hbar \begin{pmatrix} 0 & 0 & 0 \\ 1 & 0 & 0 \\ 0 & 1 & 0 \end{pmatrix}, \tag{2}
$$

where

$$
s_+ := s_x + is_y, \qquad s_- := s_x - is_y. \tag{3}
$$

An example of a spin-1 particle is the photon.

Furthermore we introduce the two matrices

$$
U = \frac{1}{\sqrt{2}} \begin{pmatrix} -1 & 0 & 1 \\ -i & 0 & -i \\ 0 & \sqrt{2} & 0 \end{pmatrix}, \qquad U^\dagger = \frac{1}{\sqrt{2}} \begin{pmatrix} -1 & i & 0 \\ 0 & 0 & \sqrt{2} \\ 1 & i & 0 \end{pmatrix}, \tag{4}
$$

where $U^\dagger$ is the transpose and complex conjugate of $U$.

In the program we evaluate

$$
\tilde{s}_x := U s_x U^\dagger \tag{5}
$$

and then the eigenvalues and eigenvectors of the matrix $\tilde{s}_x$. Furthermore we calculate

$$
s^2 := s_x^2 + s_y^2 + s_z^2. \tag{6}
$$

We find that $s^2$ is the $3 \times 3$ unit matrix times $2\hbar^2$. Obviously, $s^2$ commutes with $s_x$, $s_y$ and $s_z$.

```
%spin1.red;

matrix sp(3,3), sm(3,3), sx(3,3), sy(3,3), sz(3,3), s2(3,3);
matrix UT(3,3), U(3,3);

sp := sqrt(2)*hb*mat((0,1,0),(0,0,1),(0,0,0));
sm := sqrt(2)*hb*mat((0,0,0),(1,0,0),(0,1,0));
sx := hb/(sqrt(2))*mat((0,1,0),(1,0,1),(0,1,0));
sy := hb/(sqrt(2))*mat((0,-i,0),(i,0,-i),(0,i,0));
sz := hb*mat((1,0,0),(0,0,0),(0,0,-1));
U := 1/(sqrt(2))*mat((-1,0,1),(-i,0,-i),(0,sqrt(2),0));
Ud := sub(i=-i,tp(U));

sxp := U*sx*Ud;
mateigen(sxp,eta);

s2 := sx*sx + sy*sy + sz*sz;
```

The output is

```
UD := MAT(((  - 1)/SQRT(2),I/SQRT(2),0),(0,0,1),(1/SQRT(2),I/SQRT(2),0))$

SXP := MAT((0,0,0),(0,0, - I*HB),(0,I*HB,0))$

{{ETA,1,MAT((ARBCOMPLEX(1)),(0),(0))$},
{ - HB + ETA,1,MAT((0),( - ARBCOMPLEX(2)*I),(ARBCOMPLEX(2)))$},
{HB + ETA,1,MAT((0),(ARBCOMPLEX(3)*I),(ARBCOMPLEX(3)))$}}$

S2 := MAT((2*HB**2,0,0),(0,2*HB**2,0),(0,0,2*HB**2))$
```

## 17. Radial Symmetric Potential and Bound States

A particle of mass $m$ moves in a potential

$$V(r) = \begin{cases} -V_0 & r < a \\ 0 & r > a. \end{cases} \tag{1}$$

We determine the least value of $V_0$ such that there is a bound state of zero energy and zero angular momentum. Since there is no angular momentum, the three-dimensional eigenvalue equation reduces to

$$-\frac{\hbar^2}{2m}\left(\frac{\partial^2}{\partial r^2} + \frac{2}{r}\frac{\partial}{\partial r}\right)u = Eu, \qquad \text{outside the well} \tag{2}$$

and

$$-\frac{\hbar^2}{2m}\left(\frac{\partial^2}{\partial r^2} + \frac{2}{r}\frac{\partial}{\partial r}\right)u = (V_0 + E)u, \qquad \text{inside the well.} \tag{3}$$

With the introduction of

$$v(r) := ru(r), \tag{4}$$

the eigenvalue equations for the two domains are given by

$$\frac{d^2v}{dr^2} - \alpha^2 v = 0, \qquad r > a \tag{5}$$

and

$$\frac{d^2v}{dr^2} + \beta^2 v = 0, \qquad 0 \leq r \leq a \tag{6}$$

where

$$\alpha := \left(\frac{-2mE}{\hbar^2}\right)^{1/2}, \qquad \beta := \left(\frac{2m(E + V_0)}{\hbar^2}\right)^{1/2}. \tag{7}$$

We are interested in the limit $E \to 0^-$. Solutions of (2) and (3), respectively are given by

$$u(r) = A\frac{e^{-\alpha r}}{r}, \qquad r > a \qquad (8)$$

and

$$u(r) = B\frac{\sin(\beta r)}{r}, \qquad r < a \qquad (9)$$

where we have eliminated the solution singular at the origin. Here $A$ and $B$ are constants of integration. Continuity of $u$ and its derivative at $r = a$, or equivalently continuity of $v$ and its derivative

$$\frac{1}{v_1}\frac{dv_1}{dr} = \frac{1}{v_2}\frac{dv_2}{dr} \qquad (10)$$

requires that

$$\beta \cot(\beta a) = -\alpha. \qquad (11)$$

Here index 1 denotes the domain $r < a$ and index 2 denotes the domain $r > a$. As $E \to 0$, we find that $\alpha \to 0$. Consequently

$$\cot(\beta a) \to 0. \qquad (12)$$

This happens when

$$\beta a = \frac{\pi}{2}, \qquad (13)$$

or

$$V_0 = \frac{\pi^2 \hbar^2}{8ma^2}. \qquad (14)$$

In the program we determine $\alpha$. Thus we find (11). We set $\alpha \to alp$ and $\beta \to bet$.

40

```
%scatt3.red;

depend u1, r;
depend u2, r;
depend v1, r;
depend v2, r;

res1 := -hb*hb/(2*m)*(df(u1,r,2) + (2/r)*df(u1,r)) - en*u1;
res2 := -hb*hb/(2*m)*(df(u2,r,2) + (2/r)*df(u2,r)) - (V0 + en)*u2;
u1 := v1/r;
u2 := v2/r;
res3 := num(res1);
res4 := num(res2);
v1 := c1*exp(-alp*r);
v2 := c2*sin(bet*r);
on div;
res5 := res3/exp(-alph*r);
res6 := res4/sin(bet*r);

list1 := solve(res5,alp);
list2 := solve(res6,bet);

v1d := df(v1,r);
v2d := df(v2,r);

v1a := sub(r=a,v1);
v2a := sub(r=a,v2);

v1da := sub(r=a,v1d);
v2da := sub(r=a,v2d);

res7 := v1da/v1a - v2da/v2a;
list3 := solve(res7,alp);

part(list3,1);
alp := part(ws,2);
```

The output is

```
RES1:=(-DF(U1,R,2)*R*HB**2-2*DF(U1,R)*HB**2-2*M*R*U1*EN)/(2*M*R)$

RES2 := ( - DF(U2,R,2)*R*HB**2 - 2*DF(U2,R)*HB**2 - 2*M*R*U2*EN -
2*M*R*U2*V0)/(2*M*R)$

RES3 :=  - DF(V1,R,2)*HB**2 - 2*M*V1*EN$

RES4 :=  - DF(V2,R,2)*HB**2 - 2*M*V2*EN - 2*M*V2*V0$

RES5 := E**( - R*ALP + R*ALPH)*C1*( - 2*M*EN - HB**2*ALP**2)$

RES6 := C2*( - 2*M*EN - 2*M*V0 + HB**2*BET**2)$

LIST1 := {ALP= - SQRT(EN)*SQRT(M)*SQRT(2)*I*HB**(-1),
ALP=SQRT(EN)*SQRT(M)*SQRT(2)*I*HB**(-1)}$

LIST2 := {BET= - SQRT(2*M*EN + 2*M*V0)*HB**(-1),
BET=SQRT(2*M*EN + 2*M*V0)*HB**(-1)}$

RES7 :=  - (SIN(A*BET)**(-1)*COS(A*BET)*BET + ALP)$

LIST3 := {ALP= - SIN(A*BET)**(-1)*COS(A*BET)*BET}$

ALP :=  - SIN(A*BET)**(-1)*COS(A*BET)*BET$
```

## 18. Wave Function of Hydrogen Atom I

The Hamilton operator for the hydrogen atom is given by (MKSA-system)

$$\hat{H} := -\frac{\hbar^2}{2m}\Delta - \frac{e_0^2}{4\pi\epsilon_0 r}, \tag{1}$$

where

$$\Delta := \frac{\partial^2}{\partial x^2} + \frac{\partial^2}{\partial y^2} + \frac{\partial^2}{\partial z^2} \tag{2}$$

and

$$r := \sqrt{x^2 + y^2 + z^2}. \tag{3}$$

Taking into account only the radial part of $\hat{H}$ we obtain

$$\hat{H} = -\frac{\hbar^2}{2m}\left(\frac{d^2}{dr^2} + \frac{2}{r}\frac{d}{dr}\right) - \frac{e_0^2}{4\pi\epsilon_0 r}. \tag{4}$$

In the program we insert the ansatz

$$u(r) = C\exp(-ar), \qquad a > 0 \tag{5}$$

into the eigenvalue equation $\hat{H}u = Eu$ and determine $a$ and the energy eigenvalue. Thus we determine the ground state wave function and the ground state energy of the hydrogen atom. We obtain

$$E_0 = -\frac{me_0^4}{32\pi^2\epsilon_0^2\hbar^2} = -\frac{mc_0^2\alpha^2}{2}, \tag{6}$$

where

$$\alpha := \frac{e_0^2}{4\pi\epsilon_0\hbar c_0} \tag{7}$$

is the *Sommerfeld fine structure constant*.

```
%hyd1.red;

depend u, r;
u := C*exp(-a*r);
on div;
res1 := df(u,r,2) + (2/r)*df(u,r) + (2*m)/(hb*hb)*En*u +
(2*m*q*q)/(4*pi*ep*hb*hb)*(1/r)*u;
%q is the charge e0;


res2 := res1/u;
res3 := res2*r;


res4 := coeffn(res3,r,1);
res5 := coeffn(res3,r,0);


res6 := solve({res4,res5},{a,En});
```

The output is

```
RES1 := E**(-A*R)*C*(A**2-2*A*R**(-1) +
1/2*M*Q**2*R**(-1)*PI**(-1)*EP**(-1)*HB**(-2)+2*M*HB**(-2)*EN)$
RES2 := A**2-2*A*R**(-1)+1/2*M*Q**2*R**(-1)*PI**(-1)*EP**(-1)*HB**(-2) +
2*M*HB**(-2)*EN$
RES3 := A**2*R - 2*A + 1/2*M*Q**2*PI**(-1)*EP**(-1)*HB**(-2) +
2*M*R*HB**(-2)*EN$
RES4 := A**2 + 2*M*HB**(-2)*EN$
RES5 :=  - 2*A + 1/2*M*Q**2*PI**(-1)*EP**(-1)*HB**(-2)$
RES6 := {{A=1/4*M*Q**2*PI**(-1)*EP**(-1)*HB**(-2),

EN= - 1/32*M*Q**4*PI**(-2)*EP**(-2)*HB**(-2)}}$
```

## 19. Wave Function of Hydrogen Atom II

The eigenvalue equation for the hydrogen atom is given by

$$\left(-\frac{\hbar^2}{2m}\Delta - \frac{e_0^2}{4\pi\epsilon_0 r}\right)u(\mathbf{r}) = Eu(\mathbf{r}) \tag{1}$$

or

$$\left(\Delta + \frac{2mc_0\alpha}{\hbar r}\right)u(\mathbf{r}) = -\frac{2mE}{\hbar^2}u(\mathbf{r}), \tag{2}$$

where

$$\Delta := \frac{\partial^2}{\partial r^2} + \frac{2}{r}\frac{\partial}{\partial r} + \frac{1}{r^2\sin^2\theta}\frac{\partial^2}{\partial\phi^2} + \frac{1}{r^2}\frac{\partial^2}{\partial\theta^2} + \frac{1}{r^2}\cot\theta\frac{\partial}{\partial\theta} \tag{3}$$

and $\alpha$ is the Sommerfeld fine structure constant. Let $Y_{lm}$ be the spherical harmonics. Since

$$\left(\frac{1}{\sin^2\theta}\frac{\partial^2}{\partial\phi^2} + \frac{\partial^2}{\partial\theta^2} + \cot\theta\frac{\partial}{\partial\theta}\right)Y_{lm}(\theta,\phi) = -l(l+1)Y_{lm}(\theta,\phi) \tag{4}$$

we find that the ansatz

$$u(\mathbf{r}) = \frac{y_l(r)}{r}Y_{lm}(\theta,\phi) \tag{5}$$

is a solution of the eigenvalue equation if $y_l$ solves the *radial eigenvalue equation*

$$\frac{d^2y_l}{dr^2} + \left(\gamma^2 + \frac{2mc_0\alpha}{\hbar r} - \frac{l(l+1)}{r^2}\right)y_l = 0, \tag{6}$$

where $l$ is the orbital quantum number, $m$ is the magnetic quantum number and

$$\gamma^2 = -\frac{2mE}{\hbar^2}. \tag{7}$$

To solve this equation we insert a trial function which possesses the right asymptotic form for $r \to 0$ and $r \to \infty$ and matches the requirement of an everywhere finite and regular wave function

$$y_l(r) = r^{l+1} \exp(-\gamma r) v_l(r) \tag{8}$$

with $v_l$ a function which does not change the asymptotic feature of the ansatz.

For $v_l$ we make the polynomial ansatz

$$v_l(r) = 1 + \sum_{j=1}^{k} a_j r^j \tag{9}$$

of any chosen degree $k$ (*radial quantum number*). Then the radial eigenvalue equation reduces to a set of coupled algebraic equations, which can be solved by REDUCE to find not only the coefficients $a_j$, but also the energy eigenvalues $E_{kl}$. The results are

$$E_{kl} = -\frac{m}{2} \left( \frac{e_0^2}{4\pi\epsilon_0\hbar} \right)^2 \frac{1}{n^2} \tag{10}$$

with *main quantum number*

$$n := k + l + 1 \tag{11}$$

and

$$u_{klm}(r, \theta, \phi) = C r^l \exp(-\gamma r)\,_1F_1(l + 1 - n, 2l + 2; 2\gamma r) Y_{lm}(\theta, \phi), \tag{12}$$

where $_1F_1$ is the hypergeometric function, $n = 0, 1, 2, \ldots$ and

$$\gamma^2 = -\frac{2mE_{kl}}{\hbar^2}. \tag{13}$$

In the program we consider the case $k = 4$, $l = 1$, and $m = 0$. Thus $n = 6$.

```
%hyd2.red;

l := 1$
m := 0$
k := 4$

alp := e0^2/c/hb;

operator a$

vl := 1 + for i:=1 step 1 until k sum a(i)*r^i$

%vl is a polynomial of r of degree k with coefficients a(i);

yl := r^(l+1)*exp(-sqrt(gam2)*r)*vl$

%yl is the trial function of the radial wave function;

left := df(yl,r,2)+(-gam2 + m0*c*alp/(2*pi*hb*eps*r)-l*(l+1)/r^2)*yl;
%left is the left hand side of equation (6);

h := coeff(num(left/r),r)$
write h;

%h contains the coefficients of the power of r with "left";
%The coefficient 0 of r^0 is suppressed by num(RES/r);
%h is a set of k+1 nonlinear equations, since each coefficient
%is set to zero in order to satisfy equation (6);

b := gam2 . for i:=1 step 1 until k join {a(i)}$
write b;
%b is a list of the unknowns in the system described by h;

s := solve(h,b);
%s contains five solutions of the system, we select the first$

eigen_state := Ylm*sub(first(s),yl);

eigen_energy := - hb^2*sub(first(s),gam2)/2/m0;
```

The output is

```
left := (r*( - 24*sqrt(gam2)*a(4)*r**4*pi*eps*hb**2 +
a(4)*r**4*e0**2*m0 + 56*a(4)*r**3*pi*eps*hb**2 -
20*sqrt(gam2)*a(3)*r**3*pi*eps*hb**2 + a(3)*r**3*e0**2*m0 +
36*a(3)*r**2*pi*eps*hb**2 - 16*sqrt(gam2)*a(2)*r**2*pi*eps*hb**2 +
a(2)*r**2*e0**2*m0 + 20*a(2)*r*pi*eps*hb**2 -
12*sqrt(gam2)*a(1)*r*pi*eps*hb**2 + a(1)*r*e0**2*m0 +
8*a(1)*pi*eps*hb**2 - 8*sqrt(gam2)*pi*eps*hb**2 +
e0**2*m0))/(2*e**(sqrt(gam2)*r)*pi*eps*hb**2)$

{8*a(1)*pi*eps*hb**2-8*sqrt(gam2)*pi*eps*hb**2+
e0**2*m0,20*a(2)*pi*eps*hb**2-12*sqrt(gam2)*a(1)*pi*eps*hb**2+
a(1)*e0**2*m0,36*a(3)*pi*eps*hb**2-16*sqrt(gam2)*a(2)*pi*eps*hb**2+
a(2)*e0**2*m0,56*a(4)*pi*eps*hb**2-20*sqrt(gam2)*a(3)*pi*eps*hb**2+
a(3)*e0**2*m0,a(4)*(-24*sqrt(gam2)*pi*eps*hb**2+e0**2*m0)}$

{gam2,a(1),a(2),a(3),a(4)}$

s := {{a(4)=(e0**8*m0**4)/(17418240*pi**4*eps**4*hb**8),
a(3)=( - e0**6*m0**3)/(51840*pi**3*eps**3*hb**6),
a(2)=(e0**4*m0**2)/(480*pi**2*eps**2*hb**4),
a(1)=( - e0**2*m0)/(12*pi*eps*hb**2),
gam2=(e0**4*m0**2)/(576*pi**2*eps**2*hb**4)},
{a(4)=0,a(3)=( - e0**6*m0**3)/(120000*pi**3*eps**3*hb**6),
a(2)=(3*e0**4*m0**2)/(2000*pi**2*eps**2*hb**4),
a(1)=( - 3*e0**2*m0)/(40*pi*eps*hb**2),
gam2=(e0**4*m0**2)/(400*pi**2*eps**2*hb**4)},
{a(4)=0,a(3)=0,a(2)=(e0**4*m0**2)/(1280*pi**2*eps**2*hb**4),
a(1)=( - e0**2*m0)/(16*pi*eps*hb**2),
gam2=(e0**4*m0**2)/(256*pi**2*eps**2*hb**4)},
{a(4)=0,a(3)=0,a(2)=0,a(1)=( - e0**2*m0)/(24*pi*eps*hb**2),
gam2=(e0**4*m0**2)/(144*pi**2*eps**2*hb**4)},
{a(4)=0,a(3)=0,a(2)=0,a(1)=0,
gam2=(e0**4*m0**2)/(64*pi**2*eps**2*hb**4)}}$

eigen_state := (r**2*ylm*(r**4*e0**8*m0**4 -
336*r**3*pi*eps*e0**6*hb**2*m0**3 +
36288*r**2*pi**2*eps**2*e0**4*hb**4*m0**2 -
1451520*r*pi**3*eps**3*e0**2*hb**6*m0 +
17418240*pi**4*eps**4*hb**8))/
(17418240*e**((r*e0**2*m0)/(24*pi*eps*hb**2))*pi**4*eps**4*hb**8)$

eigen_energy := ( - e0**4*m0)/(1152*pi**2*eps**2*hb**2)$
```

## 20. Helium Atom and Trial Function

The Hamilton operator $\hat{H}$ of the Helium atom is given by

$$\hat{H} = -\frac{\hbar^2}{2m_e}(\Delta_1 + \Delta_2) + \frac{e^2}{4\pi\epsilon_0}\left(\frac{1}{r_{12}} - \frac{2}{r_1} - \frac{2}{r_2}\right) \tag{1}$$

where

$$r_{12} := |\mathbf{r}_1 - \mathbf{r}_2| \equiv \sqrt{(x_1 - x_2)^2 + (y_1 - y_2)^2 + (z_1 - z_2)^2} \tag{2}$$

with $\mathbf{r}_1 = (x_1, y_1, z_1)$ and $\mathbf{r}_2 = (x_2, y_2, z_2)$. The underlying Hilbert space is $L_2(\mathcal{R}^6)$. The eigenvalue problem cannot be solved exactly. We would like to find the ground state in an approximative manner with the help of a *variational principle*.

Let $\hat{H}$ be a Hamilton operator with discrete spectrum. Assume that the spectrum is bounded from below. Denote the lowest eigenvalue by $E_0$. Let $u \in \mathcal{H}$ where $\mathcal{H}$ is the underlying Hilbert space with $u \neq 0$. Then

$$E_0 \leq \frac{\langle u, \hat{H}u \rangle}{\langle u, u \rangle}. \tag{3}$$

If $u$ is normalized, then $E_0 \leq \langle u, \hat{H}u \rangle$. Now we give the proof of (3). Let $\{u_n : n \in I\}$ be the eigenfunctions of $\hat{H}$. Assume that $\{u_n : n \in I\}$ forms an orthonormal basis in the underlying Hilbert space $\mathcal{H}$. Since

$$u = \sum_{n \in I} c_n u_n, \qquad \hat{H}u_n = E_n u_n, \qquad \langle u_n, u_m \rangle = \delta_{nm} \tag{4}$$

we obtain

$$\langle u, \hat{H}u \rangle = \sum_{m \in I}\sum_{n \in I} c_m^* c_n E_n \langle u_m, u_n \rangle = \sum_{m \in I}\sum_{n \in I} c_m^* c_n E_n \delta_{mn} = \sum_{m \in I} c_m^* c_m E_m = \sum_{m \in I} |c_m|^2 E_m. \tag{5}$$

Since

$$\sum_{m \in I} |c_m|^2 E_m \geq \sum_{m \in I} |c_m|^2 E_0 \tag{6}$$

we obtain

$$\langle u, \hat{H}u \rangle \geq E_0 \sum_{m \in I} |c_m|^2 = E_0 \langle u, u \rangle. \tag{7}$$

Consequently, inequality (3) follows.

For the Helium atom we make the product ansatz

$$u_\mu(\mathbf{r}_1, \mathbf{r}_2) := v_\mu(r_1) v_\mu(r_2), \tag{8}$$

where

$$v_\mu(r) = \left( \frac{\mu^3}{\pi} \right)^{\frac{1}{2}} e^{\mu r}, \qquad \text{type of ground state of the } hydrogen\text{-atom}$$

This means the wave function $u_\mu \in L_2(\mathcal{R}^6)$ has only a radial part. Here $\mu$ is a real parameter. The parameter $\mu$ is determined so that

$$\frac{\langle u_\mu(r_1, r_2), \hat{H}u_\mu(r_1, r_2) \rangle}{\langle u_\mu(r_1, r_2), u_\mu(r_1, r_2) \rangle} \tag{9}$$

becomes a minimum. The calculation of the expectation values with the wave function (5) yields

$$\langle r_1^{-1} \rangle = \langle r_2^{-1} \rangle = \int_{\mathcal{R}^3} u_\mu^2(r) r^{-1} \sin\theta d\theta d\phi dr = \mu \tag{10}$$

$$\langle -\Delta_1 \rangle = \langle -\Delta_2 \rangle = -\int_{\mathcal{R}^3} u_\mu(r) \left( \frac{d^2}{dr^2} + \frac{2}{r}\frac{d}{dr} \right) u_\mu(r) \sin\theta d\theta d\phi dr = \mu^2 \tag{11}$$

$$\langle r_{12}^{-1} \rangle = 8\mu^3 \int_0^\infty \int_0^\infty \int_{-1}^1 \frac{e^{-2\mu(r_1+r_2)}}{\sqrt{r_1^2 + r_2^2 - 2r_1r_2x}} r_1^2 dr_1 r_2^2 dr_2 dx = \frac{5\mu}{8}. \tag{12}$$

We made use of the fact that in spherical coordinates we can write

$$|\mathbf{r}_1 - \mathbf{r}_2| = \sqrt{r_1^2 - 2r_1r_2\cos\theta + r_2^2} \tag{13}$$

and $x = \cos\theta$. Therefore $dx = -\sin\theta d\theta$. Consequently,

$$\frac{\langle u_\mu(r_1, r_2), \hat{H} u_\mu(r_1, r_2)\rangle}{\langle u_\mu(r_1, r_2), u_\mu(r_1, r_2)\rangle} = \frac{\hbar^2 \mu^2}{m} - \frac{27}{8}\frac{e^2}{4\pi\epsilon_0}\mu. \tag{14}$$

Minimalizing (14) with respect to $\mu$ yields

$$\mu = \mu_0 = \frac{27}{16}\frac{e^2}{4\pi\epsilon_0}\frac{m}{\hbar^2} = \frac{27}{16}\frac{m}{m_e}\frac{1}{a_0}, \tag{15}$$

where

$$m = \frac{m_e m_N}{m_e + m_N}. \tag{16}$$

Consequently

$$\frac{\langle u_{\mu 0}, \hat{H} u_{\mu 0}\rangle}{\langle u_{\mu 0}, u_{\mu 0}\rangle} = -\left(\frac{27}{16}\right)^2 \alpha^2 m c_0^2 = -77.4767\text{eV} \tag{17}$$

where $\alpha := 1/137.0381$ is the Sommerfeld fine structure constant.

In the program we evaluate

$$\int_{x=-1}^{x=1} \frac{dx}{\sqrt{r_1^2 + r_2^2 - 2r_1 r_2 x}} \tag{18}$$

and

$$\int_0^\infty r_1 e^{-2\mu r_1} dr_1 \left(2\int_0^{r_1} r_2^2 e^{-2\mu r_2} dr_2 + 2r_1 \int_{r_1}^\infty r_2 e^{-2\mu r_2} dr_2\right). \tag{19}$$

```
%inthe.red;

depend g, s;
g := 1/sqrt(r1**2 + r2**2 - 2*r1*r2*s);
res1 := int(g,s);
on precise;
res2 := sub(s=1,res1) - sub(s=-1,res1);
```

The output is

```
res1 := ( - sqrt( - 2*s*r1*r2 + r1**2 + r2**2))/(r1*r2)$
res2 := ( - abs(r1 - r2) + abs(r1 + r2))/(r1*r2)$
```


```
%heint.red;
depend res1, r1;
res1 := 2*(sub(r2=r1,int(r2*r2*exp(-2*mu*r2)),r2))
-sub(r2=0,int(r2*r2*exp(-2*mu*r2)),r2)))
-2*r1*sub(r2=r1,int(r2*exp(-2*mu*r2)),r2));
res2 := -sub(r1=0,int(r1*exp(-2*mu*r1)*res1,r1));
```

The output is

```
res1 := (e**(2*r1*mu) - r1*mu - 1)/(2*e**(2*r1*mu)*mu**3)$
res2 := 5/(64*mu**5)$
shut heintl$
```

## 21. Stark Effect

The influence exerted on an atom by a homogeneous, electrostatic field is called the Stark effect. It causes shifts and broadenings of atomic levels and thereby changes for example the absorption spectrum. We consider the hydrogen atom. We have to solve the eigenvalue equation of a hydrogen atom in an electrostatic potential (with a field strength $\mathcal{E}$ parallel to $z$-axis, MKSA-system)

$$\left(-\frac{\hbar^2}{2m_e}\Delta - \frac{e_0^2}{4\pi\epsilon_0 r} + e_0\mathcal{E}z\right)u(\mathbf{r}) = Eu(\mathbf{r}) \tag{1}$$

or

$$\left(\Delta + \frac{2m_e e_0^2}{4\pi\epsilon_0\hbar^2 r} - \frac{2m_e e_0\mathcal{E}z}{\hbar^2} + \frac{2m_e E}{\hbar^2}\right)u(\mathbf{r}) = 0 \tag{2}$$

where $\mathbf{r} := (x,y,z)$, $r := \sqrt{x^2 + y^2 + z^2}$. The quantity

$$\ell := \frac{2\pi\epsilon_0\hbar^2}{m_e e_0^2} \tag{3}$$

has the dimension of a length. Equation (2) can therefore be written as

$$\left(\Delta + \frac{1}{\ell r} - \frac{4\pi\epsilon_0}{e_0\ell}\mathcal{E}z + \frac{4\pi\epsilon_0}{e_0^2\ell}E\right)u(\mathbf{r}) = 0. \tag{4}$$

We introduce the dimensionless quantities $\tilde{x} = x/\ell$, $\tilde{y} = y/\ell$ and $\tilde{z} = z/\ell$. Moreover $\tilde{u}(\tilde{\mathbf{r}}(\mathbf{r})) = u(\mathbf{r})$. Then (4) takes the form

$$\left(\frac{\partial^2}{\partial\tilde{x}^2} + \frac{\partial^2}{\partial\tilde{y}^2} + \frac{\partial^2}{\partial\tilde{z}^2} + \frac{1}{\tilde{r}} - 2\tilde{\mathcal{E}}\tilde{z} + 2\tilde{E}\right)\tilde{u}(\tilde{\mathbf{r}}) = 0, \tag{5}$$

where we have introduced the dimensionless quantities

$$\tilde{\mathcal{E}} := \frac{2\pi\epsilon_0\ell^2\mathcal{E}}{e_0}, \qquad \tilde{E} := \frac{2\pi\epsilon_0\ell E}{e_0^2}. \tag{6}$$

It is convenient to consider (5) in *paraboloidal coordinates* $\xi$, $\eta$, $\phi$, where

$$\tilde{x} = \sqrt{\xi\eta}\cos\phi, \qquad \tilde{y} = \sqrt{\xi\eta}\sin\phi, \qquad \tilde{z} = \frac{1}{2}(\xi - \eta) \tag{7a}$$

$$\xi := \tilde{r} + \tilde{z}, \qquad \eta := \tilde{r} - \tilde{z}, \qquad \phi := \arctan(\tilde{y}/\tilde{x}), \qquad \xi + \eta = 2\tilde{r} \tag{7b}$$

and

$$v(\xi(\tilde{\mathbf{r}}), \eta(\tilde{\mathbf{r}}), \phi(\tilde{\mathbf{r}})) = \tilde{u}(\tilde{\mathbf{r}}), \tag{7c}$$

where $\xi \geq 0$, $\eta \geq 0$, and $0 \leq \phi < 2\pi$ and $2\tilde{z} = \xi - \eta$. Then (5) takes the form

$$\frac{4}{\xi + \eta}\left(\frac{\partial}{\partial\xi}\left(\xi\frac{\partial}{\partial\xi}\right) + \frac{\partial}{\partial\eta}\left(\eta\frac{\partial}{\partial\eta}\right) + \frac{\xi + \eta}{4\xi\eta}\frac{\partial^2}{\partial\phi^2}\right)v + \frac{2}{\xi + \eta}v - (\xi - \eta)\tilde{\mathcal{E}}v + 2\tilde{E}v = 0. \tag{8}$$

The paraboloidal coordinates separate the wavefunction $v$ in the product form

$$v(\xi, \eta, \phi) = f_1(\xi)f_2(\eta)\exp(im\phi), \tag{9}$$

where $m$ is the magnetic quantum number. Inserting this product ansatz into (8) and separating out the terms depending on $\xi$ and $\eta$, respectively, we find

$$\frac{d}{d\xi}\left(\xi\frac{df_1}{d\xi}\right) + \left(\frac{1}{2}\tilde{E}\xi - \frac{m^2}{4\xi} - \frac{1}{4}\tilde{\mathcal{E}}\xi^2\right)f_1 = -\beta_1 f_1 \tag{10a}$$

$$\frac{d}{d\eta}\left(\eta\frac{df_1}{d\eta}\right) + \left(\frac{1}{2}\tilde{E}\eta - \frac{m^2}{4\eta} + \frac{1}{4}\tilde{\mathcal{E}}\eta^2\right)f_2 = -\beta_2 f_2, \tag{10b}$$

where $\beta_1 + \beta_2 = 2$. Since

$$f_1(\xi, \tilde{\mathcal{E}}) = f_2(\xi, -\tilde{\mathcal{E}}), \qquad \tilde{E}(\tilde{\mathcal{E}}) = \tilde{E}(-\tilde{\mathcal{E}}), \qquad \beta_1(\tilde{\mathcal{E}}) = \beta_2(-\tilde{\mathcal{E}}) \tag{11}$$

it is sufficient to consider only the first of these equations. Substituting

$$w(\xi) := -2\frac{1}{f_1}\frac{df_1}{d\xi}, \quad \text{or} \quad f_1(\xi) := \exp\left(-\frac{1}{2}\int^\xi w(s)ds\right) \tag{12}$$

into (10a) we obtain the *Riccati equation*

$$\xi\frac{dw}{d\xi} - \frac{1}{2}\xi w^2 + w - \tilde{E}\xi + \frac{m^2}{2\xi} + \frac{1}{2}\xi^2\tilde{E} = 2\beta, \tag{13}$$

where we set $\beta_1 = \beta$. Notice that

$$\frac{d(\xi w)}{d\xi} \equiv \xi\frac{dw}{d\xi} + w. \tag{14}$$

In the following we consider the case $m = 0$. This equation can be solved by expanding $\tilde{E}$, $\beta$ and $w$ into powers of the field strength $\tilde{\mathcal{E}}$

$$\tilde{E} = -\frac{1}{2}\left(1 + \sum_{k=1}^{\infty} d_k\tilde{\mathcal{E}}^k\right), \qquad \beta = \frac{1}{2}\left(1 + \sum_{k=1}^{\infty} b_k\tilde{\mathcal{E}}^k\right) \tag{15a}$$

$$w(\xi) = \sum_{k=0}^{\infty} w_k(\xi)\tilde{\mathcal{E}}^k, \qquad w_k(\xi) = (-1)^{k+1}\sum_{l=0}^{k} a_l^{(k)}\tilde{\mathcal{E}}^k. \tag{15b}$$

In our first program (Stark1.red) we insert (9) into (8) in order to obtain (10a) and (10b). In our second program (Stark2.red) we insert (12) into (10a) and obtain (13). The third program (Stark3.red) starts by forming these symbolic series up to a given power of $\tilde{\mathcal{E}}$, which are then inserted into (13). Subsequently, by comparison of the coefficients of the powers of $\tilde{\mathcal{E}}$ and $\xi$ we find the coefficients

$$d_k, \quad b_k, \quad a_l^{(k)}. \tag{16}$$

In the programs we set $\tilde{E} = en$ and $\tilde{\mathcal{E}} = ef$.

```
%stark1.red;

depend v, xi, eta, phi;
depend f1, xi;
depend f2, eta;

%product ansatz;
v := f1*f2*exp(i*m*phi);

%equation 8;
res1 := 4/(xi+eta)*(df(xi*df(v,xi),xi) + df(eta*df(v,eta),eta) +
(xi+eta)/(4*xi*eta)*df(v,phi,2)) + 2/(xi+eta)*v -
(xi-eta)*ef*v + 2*en*v;

on div;
res2 := res1/(f1*f2*exp(i*m*phi));
```

The output is

```
res2 := (4*df(f1,xi,2)*xi*f1**(-1) + 4*df(f1,xi)*f1**(-1) +
4*df(f2,eta,2)*eta*f2**(-1) + 4*df(f2,eta)*f2**(-1) -
m**2*xi**(-1) - m**2*eta**(-1) - xi**2*ef + 2*xi*en +
eta**2*ef + 2*eta*en + 2)/(xi + eta)$
```

*Remark:* Separating out *res2* with respect to $\xi$ and $\eta$ yields (10a) and (10b).

```
%stark2.red;

let df(temp,xi) = w;

depend w, xi;

f := exp(-1/2*temp);

res := df(f,xi) + xi*df(f,xi,2) +
(1/2*en*xi - m2/(4*xi) - 1/4*ef*xi**2 + beta)*f;

res := res/f;

res := -2*res;

on div;

res;
```

The output is

```
res := ( - 2*df(w,xi)*xi**2 + w**2*xi**2 - 2*w*xi - xi**3*ef +
2*xi**2*en + 4*xi*beta - m2)/(4*e**(temp/2)*xi)$

res := ( - 2*df(w,xi)*xi**2 + w**2*xi**2 - 2*w*xi - xi**3*ef +
2*xi**2*en + 4*xi*beta - m2)/(4*xi)$

res := (2*df(w,xi)*xi**2 - w**2*xi**2 + 2*w*xi + xi**3*ef -
2*xi**2*en - 4*xi*beta + m2)/(2*xi)$

df(w,xi)*xi - 1/2*w**2*xi + w + 1/2*xi**2*ef - xi*en +
1/2*xi**(-1)*m2 - 2*beta$
```

```
%stark3.red;
%kmax determines the highest order of the
%perturbation series in the field strength;

define kmax = 4;
operator a, b, d;
array g(kmax+1), h(kmax);
for k := 0 step 2 until kmax do <<b(k) := 0; d(k+1) := 0>>;

% Polynomial ansatz for w, beta and energy;
w := 1 - for k := 1 step 1 until kmax sum
     (-ef)^k*(for l := 0 step 1 until k sum a(l,k)*xi^l);
en := -1/2*(1 + for k := 1 step 1 until kmax sum d(k)*ef^k);
beta := 1/2*(1 + for k := 1 step 1 until kmax sum b(k)*ef^k);

let ef^(kmax + 1) = 0;
u := df(xi*w,xi)-1/2*xi*w^2-en*xi+1/2*ef*xi^2-2*beta;
h := coeff(num(u),ef);

g := for k:=1 step 1 until length(h) join coeff(part(h,k),xi)$
s := for k:=0 step 1 until kmax join
     for l:=k step 1 until kmax join {a(k,l)}$

for k:=0 step 2 until kmax do s := append(s,{b(k+1),d(k)})$

b := solve(g,s);
en := sub(b,en);
```

The output is

```
b := {{a(0,1)=1,a(1,1)=1/2,a(0,2)=0,a(1,2)=7/8,a(2,2)=1/8,
a(3,3)=1/16,a(2,3)=13/16,a(1,3)=53/16,a(0,3)=53/8,
a(4,4)=5/128,a(3,4)=99/128,a(2,4)=761/128,a(1,4)=3131/128,
a(0,4)=0,
d(4)=3555/32,d(2)=9/2,
b(3)=53/8,b(1)=1}}$


en := ( - 3555*ef**4 - 144*ef**2 - 32)/64$
```

## 22. Scattering in One-Dimension

We calculate the transmission and reflection coefficients of a particle having total energy $E$, at the potential barrier given by

$$
V(q) = \begin{cases} 0 & \text{if} \quad q < 0 \\ V_0 & \text{if} \quad 0 \leq q \leq d \\ 0 & \text{if} \quad q > d \end{cases} \tag{1}
$$

where $V_0 > 0$ and $E > V_0$. The general solution of the eigenvalue equation

$$
\hat{H}u(q) = Eu(q) \tag{2}
$$

in the three domains is given by

$$
u(q) = \begin{cases} Ae^{ip_1 q/\hbar} + Be^{-ip_1 q/\hbar} & q < 0 \\ Ge^{ip_2 q/\hbar} + Fe^{-ip_2 q/\hbar} & 0 \leq q \leq d \\ Ce^{ip_1 q/\hbar} & q > d \end{cases} \tag{3}
$$

where

$$
p_1 = \sqrt{2mE}, \qquad p_2 = \sqrt{2m(E - V_0)}. \tag{4}
$$

If all particles arrive at the barrier from the left, the terms with the coefficients $A$, $B$ and $C$ represent, respectively, the incident, the reflected and the transmitted wave. Obviously, for $q > d$ there is no reflected wave.

The continuity condition for $u$ at $q = 0$ and at $q = d$ gives

$$
A + B = G + F \tag{5a}
$$

$$
Ge^{ip_2 d/\hbar} + Fe^{-ip_2 d/\hbar} = Ce^{ip_1 d/\hbar}. \tag{5b}
$$

The continuity condition for $du/dq$ at $q = 0$ and at $q = d$ gives

$$
p_1(A - B) = p_2(G - F) \tag{5c}
$$

$$
p_2(Ge^{ip_2 d/\hbar} - Fe^{-ip_2 d/\hbar}) = p_1 Ce^{ip_1 d/\hbar}. \tag{5d}
$$

Eliminating $G$, $F$ and $C$ we obtain $B/A$. The reflection coefficient is given by

$$R = \left|\frac{B}{A}\right|^2.$$

(6)

In the program we evaluate $B/A$ for the case $E > V_0$. We find that

$$\frac{B}{A} = \frac{(p_1^2 - p_2^2)(1 - e^{2ip_2a/\hbar})}{(p_1 + p_2)^2 - (p_1 - p_2)^2 e^{2ip_2a/\hbar}}.$$

(7)

*Exercise:* Solve the case for $E < V_0$.

60

```
%scatt.red;

operator ul, um, ur, ul1, um1, ur1;

depend ul, q;
depend um, q;
depend ur, q;

ul := A*exp(i*p1*q/hb) + B*exp(-i*p1*q/hb);
um := G*exp(i*p2*q/hb) + F*exp(-i*p2*q/hb);
ur := C*exp(i*p1*q/hb);

ul1 := df(ul,q);
um1 := df(um,q);
ur1 := df(ur,q);

B1 := sub(q=0,ul) - sub(q=0,um);
B2 := sub(q=d,um) - sub(q=d,ur);
B3 := sub(q=0,ul1) - sub(q=0,um1);
B4 := sub(q=d,um1) - sub(q=d,ur1);

L1 := solve(B1=0,F);
F := part(part(L1,1),2);
B2;
B3;
B4;
L2 := solve(B2,G);
G := part(part(L2,1),2);
B3; B4;
L3 := solve(B3,C);
C := part(part(L3,1),2);
B4;
L4 := solve(B4,A);
A := part(part(L4,1),2);

RES1 := B/A;
```

The output is

```
ul1 := (i*p1*(e**((2*i*q*p1)/hb)*a - b))/(e**((i*q*p1)/hb)*hb)$
um1 := (i*p2*(e**((2*i*q*p2)/hb)*g - f))/(e**((i*q*p2)/hb)*hb)$

b1 := a + b - f - g$
b2 := (-e**((d*i*p1 + d*i*p2)/hb)*c + e**((2*d*i*p2)/hb)*g+ f)/
e**((d*i*p2)/hb)$
b3 := (i*(a*p1 - b*p1 + f*p2 - g*p2))/hb$
b4 := (i*( - e**((d*i*p1 + d*i*p2)/hb)*c*p1 + e**((2*d*i*p2)/
hb)*g*p2 - f*p2))/(e**((d*i*p2)/hb)*hb)$

f := a + b - g$

l2 := {g=(e**((d*i*p1 + d*i*p2)/hb)*c - a - b)/(e**((2*d*i*p2)/
hb) - 1)}$

g := (e**((d*i*(p1 + p2))/hb)*c - a - b)/(e**((2*d*i*p2)/hb)- 1)$

l3 := {c=(e**((2*d*i*p2)/hb)*a*p1 + e**((2*d*i*p2)/hb)*a*p2 -
 e**((2*d*i*p2)/hb)*b*p1 + e**((2*d*i*p2)/hb)*b*p2
- a*p1 + a*p2 + b*p1 + b*p2)/(2*e**((d*i*p1 + d*i*p2)/hb)*p2)}$

c := (e**((2*d*i*p2)/hb)*a*p1 + e**((2*d*i*p2)/hb)*a*p2 -
e**((2*d*i*p2)/hb)*b*p1 + e**((2*d*i*p2)/hb)*b*p2 - a*p1 +
 a*p2 + b*p1 + b*p2)/(2*e**((d*i*(p1 + p2))/hb)*p2)$

l4 := {a=(b*(e**((2*d*i*p2)/hb)*p1**2 - 2*e**((2*d*i*p2)/hb)*p1*p2 +
e**((2*d*i*p2)/hb)*p2**2 - p1**2 - 2*p1*p2 - p2**2))/
(e**((2*d*i*p2)/hb)*p1**2 - e**((2*d*i*p2)/hb)*p2**2 -
p1**2 + p2**2)}$

a := (b*(e**((2*d*i*p2)/hb)*p1**2 - 2*e**((2*d*i*p2)/hb)*p1*p2 +
e**((2*d*i*p2)/hb)*p2**2 - p1**2 - 2*p1*p2 - p2**2))/
(e**((2*d*i*p2)/hb)*p1**2 - e**((2*d*i*p2)/hb)*p2**2 -
p1**2 + p2**2)$

res1 := (e**((2*d*i*p2)/hb)*p1**2 - e**((2*d*i*p2)/hb)*p2**2
- p1**2 + p2**2)/
(e**((2*d*i*p2)/hb)*p1**2 - 2*e**((2*d*i*p2)/hb)*p1*p2 +
e**((2*d*i*p2)/hb)*p2**2 - p1**2 - 2*p1*p2 - p2**2)$
```

## 23. Gauge Theory

Let

$$i\hbar\frac{\partial\psi(\mathbf{r},t)}{\partial t} = -\frac{\hbar^2}{2m}\Delta\psi(\mathbf{r},t) \tag{1}$$

be the Schrödinger equation of the free particle in three space dimensions. Equation (1) is invariant under the transformation

$$\psi'(\mathbf{r}'(\mathbf{r},t),t'(\mathbf{r},t)) = \exp(i\epsilon)\psi(\mathbf{r},t) \tag{2a}$$

$$x_j'(\mathbf{r},t) = x_j, \qquad t'(\mathbf{r},t) = t, \tag{2b}$$

where $j = 1,2,3$. Transformation (2) is called a *global gauge transformation*. Let

$$i\hbar\frac{\partial\psi}{\partial t} = -\frac{\hbar^2}{2m}\sum_{j=1}^{3}\left(\frac{\partial}{\partial x_j} - i\frac{q}{\hbar}A_j\right)^2\psi + qU\psi, \tag{3}$$

where $\mathbf{A} = (A_1, A_2, A_3)$ is the vector potential and $U$ the scalar potential. We show that (3) is invariant under the transformation

$$x_j'(\mathbf{r},t) = x_j, \qquad t'(\mathbf{r},t) = t \tag{4a}$$

$$\psi'(\mathbf{r}'(\mathbf{r},t),t'(\mathbf{r},t)) = \exp(i\epsilon(\mathbf{r},t))\psi(\mathbf{r},t) \tag{4b}$$

$$U'(\mathbf{r}'(\mathbf{r},t),t'(\mathbf{r},t)) = U(\mathbf{r},t) - \frac{\hbar}{q}\frac{\partial\epsilon(\mathbf{r},t)}{\partial t} \tag{4c}$$

$$A_j'(\mathbf{r}'(\mathbf{r},t),t'(\mathbf{r},t)) = A_j(\mathbf{r},t) + \frac{\hbar}{q}\frac{\partial\epsilon(\mathbf{r},t)}{\partial x_j}, \tag{4d}$$

where $j = 1,2,3$ and $q$ is the charge. This transformation is called a *local gauge transformation*. We have to show that

$$i\hbar\frac{\partial\psi'}{\partial t'} = -\frac{\hbar^2}{2m}\sum_{j=1}^{3}\left(\frac{\partial}{\partial x_j'} - i\frac{q}{\hbar}A_j'\right)^2\psi' + qU'\psi'. \tag{5}$$

```
%gauge.red;

depend GP, x1, x2, x3, t;
depend G, x1, x2, x3, t;
depend psip, x1, x2, x3, t;
depend psi, x1, x2, x3, t;
depend AP1, x1, x2, x3, t;
depend AP2, x1, x2, x3, t;
depend AP3, x1, x2, x3, t;
depend A1, x1, x2, x3, t;
depend A2, x1, x2, x3, t;
depend A3, x1, x2, x3, t;
depend UP, x1, x2, x3, t;
depend U, x1, x2, x3, t;
depend ep, x1, x2, x3, t;
GP := i*hb*df(psip,t) +
hb*hb/(2*m)*(df(psip,x1,2)+df(psip,x2,2)+df(psip,x3,2)) +
(-i*q*hb)/(2*m)*(AP1*df(psip,x1)+AP2*df(psip,x2)+AP3*df(psip,x3))+
(-i*q*hb)/(2*m)*(df(psip*AP1,x1)+df(psip*AP2,x2)+df(psip*AP3,x3))-
q^2/(2*m)*(AP1*AP1*psip+AP2*AP2*psip+AP3*AP3*psip)-q*UP*psip;

G := sub({psip=psi*exp(i*ep),AP1=A1+hb/q*df(ep,x1),
AP2=A2+hb/q*df(ep,x2),AP3=A3+hb/q*df(ep,x3),UP=U-hb/q*df(ep,t)},GP);

on div;
G1 := G/EXP(i*ep);
G2 := sub({psi=psip,A1=AP1,A2=AP2,A3=AP3,U=UP},G1);
RES := GP - G2;
```

The output is

```
g1 := - 1/2*df(a1,x1)*i*m**(-1)*q*psi*hb -
1/2*df(a2,x2)*i*m**(-1)*q*psi*hb + df(psi,t)*i*hb +
1/2*df(psi,x1,2)*m**(-1)*hb**2 - df(psi,x1)*i*m**(-1)*q*a1*hb +
1/2*df(psi,x2,2)*m**(-1)*hb**2 - df(psi,x2)*i*m**(-1)*q*a2*hb +
1/2*df(psi,x3,2)*m**(-1)*hb**2 - df(psi,x3)*i*m**(-1)*q*a3*hb -
1/2*df(a3,x3)*i*m**(-1)*q*psi*hb - 1/2*m**(-1)*q**2*a1**2*psi -
1/2*m**(-1)*q**2*a2**2*psi - 1/2*m**(-1)*q**2*psi*a3**2 -
q*u*psi$

res := 0$
```

## 24. Driven Two Level System

Consider an atomic system with two stationary states $|1\rangle$ and $|2\rangle$. Let $\hat{H}$ be the Hamilton operator with

$$\hat{H}|1\rangle = \hbar\omega_1|1\rangle, \qquad \hat{H}|2\rangle = \hbar\omega_2|2\rangle. \tag{1}$$

At time $t = 0$ a periodic perturbation $W\cos(\omega t)$ is switched on (e.g. a light wave) with frequency $\omega$. We evaluate the probability of finding the atomic system in either state at time $t$. The Schrödinger equation

$$i\hbar\frac{\partial\psi}{\partial t} = (\hat{H} + W\cos(\omega t))\psi \tag{2}$$

has to be solved, where

$$\psi(t) = c_1(t)e^{-i\omega_1 t}|1\rangle + c_2(t)e^{-i\omega_2 t}|2\rangle. \tag{3}$$

The initial condition is given by

$$\psi(t = 0) = |1\rangle. \tag{4}$$

From (4) we have

$$c_1(t = 0) = 1, \qquad c_2(t = 0) = 0. \tag{5}$$

Inserting (3) into (2) and taking the scalar product with either $\langle 1|$ or $\langle 2|$ we arrive at

$$\frac{dc_1}{d\tau} = -i\cos(\tau)(\tilde{W}_{11}c_1 + \tilde{W}_{12}c_2(\cos(\tilde{\omega}_0\tau) - i\sin(\tilde{\omega}_0\tau))) \tag{6a}$$

$$\frac{dc_2}{d\tau} = -i\cos(\tau)(\tilde{W}_{21}c_1(\cos(\tilde{\omega}\tau) + i\sin(\tilde{\omega}_0\tau)) + \tilde{W}_{22}c_2), \tag{6b}$$

where $W_{ij} := \langle i|W|j\rangle$ $(W_{ij} = W_{ji})$ and

$$\tau := \omega t, \qquad \omega_0 := \omega_2 - \omega_1, \qquad \tilde{\omega}_0 := \frac{\omega_0}{\omega}, \qquad \tilde{W}_{ij} = \frac{W_{ij}}{\hbar\omega}. \tag{7}$$

We solve (6) numerically with the Lie series technique and evaluate the probabilities

$$|c_1(\tau)|^2, \qquad |c_2(\tau)|^2, \qquad |c_1(\tau)|^2 + |c_2(\tau)|^2 = 1. \tag{8}$$

We have $c_1 = c_{1r} + ic_{1c}$, $c_2 = c_{2r} + ic_{rc}$ and set $c_{1r} = x(1)$, $c_{1i} = x(2)$, $c_{2r} = x(3)$, $c_{2i} = x(4)$, $\tau = x(5)$.

```
%two.red;

operator V, Q, x, xs;
depend V(j), x(k);  depend Q(j), x(k);

V(1) := cos(x(5))*w11*x(2)+cos(x(5))*cos(om*x(5))*w12*x(4)
-cos(x(5))*sin(om*x(5))*w12*x(3);
V(2) := -cos(x(5))*w11*x(1)-cos(x(5))*cos(om*x(5))*w12*x(3)
-cos(x(5))*sin(om*x(5))*w12*x(4);
V(3) := cos(x(5))*cos(om*x(5))*w21*x(2)+cos(x(5))*sin(om*x(5))*w21*x(1)
+cos(x(5))*w22*x(4);
V(4) := -cos(x(5))*cos(om*x(5))*w21*x(1)+cos(x(5))*sin(om*x(5))*w21*x(2)
-cos(x(5))*w22*x(3);
V(5) := 1;

for j:=1:5 do
Q(j) := for k := 1:5 sum (V(k)*df(V(j),x(k)));

on rounded$
om := 0.1; w11 := 0.5; w12 := 0.5; w21 := 0.5; w22 := 0.5;
count := 0.0; ep := 0.01;                    % ep : step length;
xs(1) := 1.0; xs(2) := 0.0; xs(3) := 0.0; % initial value
xs(4) := 0.0; xs(5) := 0.0;

while count < 20 do
<< x(1) := xs(1); x(2) := xs(2); x(3) := xs(3);
x(4) := xs(4); x(5) := xs(5);
xs(1) := x(1) + ep*V(1) + ((ep*ep)/2)*Q(1);
xs(2) := x(2) + ep*V(2) + ((ep*ep)/2)*Q(2);
xs(3) := x(3) + ep*V(3) + ((ep*ep)/2)*Q(3);
xs(4) := x(4) + ep*V(4) + ((ep*ep)/2)*Q(4);
xs(5) := x(5) + ep*V(5) + ((ep*ep)/2)*Q(5);
count := count + ep>>$
write "count = ", count;
prob1:=xs(1)*xs(1)+xs(2)*xs(2); prob2:=xs(3)*xs(3)+xs(4)*xs(4);
```

The output is

```
count = 20.0$  prob1 := 0.83722170531$  prob2 := 0.16275975343$
```

## 25. Free Electron Spin Resonance

A free electron is put inside a cavity in which there are two magnetic fields, viz. a constant homogeneous field $\mathbf{B}$ in $z$-direction and a field $\mathbf{B'}$ rotating in the $x, y$ plane. Thus the components of the fields are given by

$$B_x = 0, \qquad B_y = 0, \qquad B_z = B_0 \tag{1}$$

$$B'_x = B' \cos(\omega t), \qquad B'_y = B' \sin(\omega t), \qquad B'_z = 0. \tag{2}$$

The Hamilton operator (time-dependent $2 \times 2$ matrix) is given by

$$\hat{H} = \mu_B(\sigma_z B_0 + \sigma_x B'_x + \sigma_y B'_y), \tag{3}$$

where $\mu_B := e_0 \hbar / 2m$ and $\sigma_x$, $\sigma_y$ and $\sigma_z$ are the Pauli spin matrices. The Schrödinger equation is given by

$$i\hbar \frac{\partial \Psi}{\partial t} = \mu_B \left( B_0 \sigma_z + \frac{1}{2} B'(e^{-i\omega t}\sigma_+ + e^{i\omega t}\sigma_-) \right)\Psi, \tag{4}$$

where

$$\Psi(t) = \begin{pmatrix} u(t) \\ v(t) \end{pmatrix} \tag{5}$$

and $\sigma_+ := \sigma_x + i\sigma_y$, $\sigma_- := \sigma_x - i\sigma_y$. Let

$$\omega_0 := \frac{\mu_B B_0}{\hbar}, \qquad \omega' := \frac{\mu_B B'}{\hbar}. \tag{6}$$

Then the Schrödinger equation (4) takes the form

$$i\frac{du}{dt} = \omega_0 u + \omega' e^{-i\omega t} v \tag{7a}$$

$$i\frac{dv}{dt} = -\omega_0 v + \omega' e^{i\omega t} u. \tag{7b}$$

This linear system of differential equations can be solved by

$$u(t) = Ae^{-i(\Omega + \frac{1}{2}\omega)t}, \qquad v(t) = Be^{-i(\Omega - \frac{1}{2}\omega)t}.$$ (8)

We obtain two solutions $\Omega_1 = +\Omega$ and $\Omega_2 = -\Omega$ with

$$\Omega := \sqrt{(\omega_0 - \frac{1}{2}\omega)^2 + \omega'^2}.$$ (9)

Thus the wave function is given by

$$\Psi(t) = (A_1 e^{-i\Omega t} + A_2 e^{i\Omega t})e^{-i\omega t/2} \begin{pmatrix} 1 \\ 0 \end{pmatrix} + (B_1 e^{-i\Omega t} + B_2 e^{i\Omega t})e^{i\omega t/2} \begin{pmatrix} 0 \\ 1 \end{pmatrix},$$ (10)

where

$$B_{1,2} = A_{1,2}\frac{\pm\Omega - (\omega_0 - \frac{1}{2}\omega)}{\omega'}.$$ (11)

Assume that the initial condition is

$$\Psi(t = 0) = \begin{pmatrix} 1 \\ 0 \end{pmatrix}.$$ (12)

Then

$$A_1 + A_2 = 1, \qquad B_1 + B_2 = 0.$$ (13)

Thus we find for the solution of the Schrödinger equation (4)

$$\Psi(t) = \left(\cos(\Omega t) - \frac{\omega_0 - \omega/2}{\Omega}i\sin(\Omega t)\right)e^{-i\omega t/2}\begin{pmatrix} 1 \\ 0 \end{pmatrix} - \frac{\omega'}{\Omega}i\sin(\Omega t)e^{i\omega t/2}\begin{pmatrix} 0 \\ 1 \end{pmatrix}.$$ (14)

In the first program we insert (8) into (7) and evaluate $\Omega$. We set $\omega_0$ =om0, $\omega'$ =omp, $\omega$ =oms, and $\Omega$ =omc. In the second program we insert (14) into (7) and check that it is in fact the solution of the initial value problem.

```
%spres.red;

depend u, t;
depend v, t;

u := A*exp(-i*(omc + oms/2)*t);
v := B*exp(-i*(omc - oms/2)*t);

res1 := i*df(u,t) - om0*u - omp*exp(-i*oms*t)*v;
res1 := res1*2*exp((2*i*t*omc + i*t*oms)/2);
res2 := i*df(v,t) + om0*v - omp*exp(i*oms*t)*u;
res2 := res2*2*exp((2*i*t*omc + i*t*oms)/2)/exp(i*t*oms);

matrix m(2,2);
m := mat((coeffn(res1,A,1),coeffn(res1,B,1)),
(coeffn(res2,A,1),coeffn(res2,B,1)));

r := det(m);

list := solve(r=0,omc);
```

The output is

```
res1 := (2*a*omc + a*oms - 2*a*om0 - 2*b*omp)/
(2*e**((2*i*t*omc + i*t*oms)/2))$
res1 := 2*a*omc + a*oms - 2*a*om0 - 2*b*omp$

res2 := (e**(i*t*oms)*( - 2*a*omp + 2*b*omc - b*oms + 2*b*om0))/
(2*e**((2*i*t*omc + i*t*oms)/2))$
res2 := - 2*a*omp + 2*b*omc - b*oms + 2*b*om0$

m:=mat((2*omc+oms-2*om0,-2*omp),(-2*omp,2*omc-oms+2*om0))$
r := 4*omc**2 - oms**2 + 4*oms*om0 - 4*om0**2 - 4*omp**2$

list := {omc= - 1/2*sqrt(oms**2 - 4*oms*om0 + 4*om0**2 + 4*omp**2),
omc=1/2*sqrt(oms**2 - 4*oms*om0 + 4*om0**2 + 4*omp**2)}$
```

*Remark:* List gives the two solutions of (9).

```
%spres1.red;

depend u, t;
depend v, t;

u := (cos(omc*t)-(om0-oms/2)/omc*i*sin(omc*t))*exp(-i*oms*t/2);
v := -(omp/omc)*i*sin(omc*t)*exp(i*oms*t/2);

res1 := i*df(u,t) - om0*u - omp*exp(-i*oms*t)*v;
res2 := i*df(v,t) + om0*v - omp*exp(i*oms*t)*u;

for all q let sin(q) = (exp(i*q)-exp(-i*q))/2;
for all q let cos(q) = (exp(i*q)+exp(-i*q))/2;

res1;
res2;

let omc = sqrt((om0-oms/2)**2+omp**2);

res1;
res2;
```

The output is

```
res1 := (sin(t*omc)*i*( - 4*omc**2 + 4*om0**2 - 4*om0*oms +
oms**2 + 4*omp**2))/(4*e**((i*t*oms)/2)*omc)$

res2 := 0$

res1;
(i*( - 4*e**(2*i*t*omc)*omc**2 + 4*e**(2*i*t*omc)*om0**2 - 4*
e**(2*i*t*omc)*om0*oms + e**(2*i*t*omc)*oms**2 + 4*e**(2*
i*t*omc)*omp**2 + 4*omc**2 - 4*om0**2 + 4*om0*oms -
oms**2 - 4*omp**2))/(8*e**((2*i*t*omc + i*t*oms)/2)*omc)$

res2;
0$

res1;
0$

res2;
0$
```

# 26. Two-Point Ising-Model with External Field

We consider the two-point Ising model with an external field

$$\hat{H} = J\sigma_{z,1}\sigma_{z,2} + B(\sigma_{x,1} + \sigma_{x,2}),\qquad (1)$$

where

$$\sigma_x = \begin{pmatrix} 0 & 1 \\ 1 & 0 \end{pmatrix}, \qquad \sigma_y = \begin{pmatrix} 0 & i \\ -i & 0 \end{pmatrix}, \qquad \sigma_z = \begin{pmatrix} 1 & 0 \\ 0 & -1 \end{pmatrix} \qquad (2)$$

and $J$ and $B > 0$ are constants. We have

$$\sigma_{z,1} = \sigma_z \otimes I, \quad \sigma_{z,2} = I \otimes \sigma_z, \quad \sigma_{x,1} = \sigma_x \otimes I, \quad \sigma_{x,2} = I \otimes \sigma_x, \qquad (3)$$

where $\otimes$ denotes the Kronecker product (see below for the definition) and $I$ is the $2 \times 2$ unit matrix. Thus

$$\hat{H} = J(\sigma_z \otimes \sigma_z) + B(\sigma_x \otimes I + I \otimes \sigma_x) \qquad (4)$$

is a $4 \times 4$ matrix.

From (4) we find that the matrix $\hat{H}$ takes the form

$$\hat{H} = \begin{pmatrix} J & B & B & 0 \\ B & -J & 0 & B \\ B & 0 & -J & B \\ 0 & B & B & J \end{pmatrix}. \qquad (5)$$

The eigenvalues are

$$E_{1,2} = \pm\sqrt{J^2 + 4B^2}, \qquad E_3 = -J, \qquad E_4 = J. \qquad (6)$$

Let us now give the definition of the *Kronecker product*. Let $A$ be an $m \times n$ matrix and let $B$ be a $p \times q$ matrix. Then the Kronecker product of $A$ and $B$ is the $(mp) \times (nq)$ matrix defined by

$$A \otimes B := \begin{pmatrix} a_{11}B & a_{12}B & \cdots & a_{1n}B \\ a_{21}B & a_{22}B & \cdots & a_{2n}B \\ \vdots & & & \\ a_{m1}B & a_{m2}B & \cdots & a_{mn}B \end{pmatrix}. \qquad (7)$$

Let

$$A = \begin{pmatrix} 2 & 3 \\ 0 & 1 \end{pmatrix}, \qquad B = \begin{pmatrix} 0 & -1 \\ -1 & 1 \end{pmatrix}. \tag{8}$$

Then

$$A \otimes B = \begin{pmatrix} 0 & -2 & 0 & -3 \\ -2 & 2 & -3 & 3 \\ 0 & 0 & 0 & -1 \\ 0 & 0 & -1 & 1 \end{pmatrix}, \qquad B \otimes A = \begin{pmatrix} 0 & 0 & -2 & -3 \\ 0 & 0 & 0 & -1 \\ -2 & -3 & 2 & 3 \\ 0 & -1 & 0 & 1 \end{pmatrix}. \tag{9}$$

Note that

$$(A \otimes B)(C \otimes D) = (AC) \otimes (BD), \tag{10}$$

where $AC$ and $BD$ denote the matrix products (we assume they exist). Let $X$ be an $n \times n$ matrix and $Y$ be an $m \times m$ matrix. Then

$$\text{tr}(X \otimes Y) = \text{tr}(X)\text{tr}(Y). \tag{11}$$

In the program we implement the Kronecker product of two $n \times n$ matrices as a procedure. Then we evaluate the Hamilton operator $\hat{H}$. Finally we determine the eigenvalues and eigenvectors of $\hat{H}$.

72

```
%Program name: isingex.red;

procedure Kron(A,B);
begin
n := 2; m := n*n;

operator A$
matrix AA(n,n)$
for i:=1:n do
for j:=1:n do
AA(i,j):=A(i,j)$

operator B$
matrix BB(n,n)$
for i:=1:n do
for j:=1:n do
BB(i,j):=B(i,j)$

operator C$
matrix CC(m,m);
c1 := 0; c2 := 0;
for r:=1:n do
for s:=1:n do
begin
for i:=1:n do
for j:=1:n do
begin
c1 := n*(r-1); c2 := n*(s-1);
CC(i+c1,j+c2) := AA(r,s)*BB(i,j);
end;
end;
return CC;
end;

operator SZ, SX, ID;
SZ(1,1) := 1; SZ(1,2) := 0;
SZ(2,1) := 0; SZ(2,2) := -1;

SX(1,1) := 0; SX(1,2) := 1;
SX(2,1) := 1; SX(2,2) := 0;

ID(1,1) := 1; ID(1,2) := 0; ID(2,1) := 0; ID(2,2) := 1;
```

```
R1 := Kron(SZ,ID);
R2 := Kron(ID,SZ);
R3 := R1*R2;
R4 := Kron(SX,ID);
R5 := Kron(ID,SX);
R6 := R4 + R5;
R7 := J*R3 + B*R6;
MATEIGEN(R7,eta);
```

The output is

```
r1 := mat((1,0,0,0),(0,1,0,0),(0,0,-1,0),(0,0,0,-1))$
r2 := mat((1,0,0,0),(0,-1,0,0),(0,0,1,0),(0,0,0,-1))$
r3 := mat((1,0,0,0),(0,-1,0,0),(0,0,-1,0),(0,0,0,1))$
r4 := mat((0,0,1,0),(0,0,0,1),(1,0,0,0),(0,1,0,0))$
r5 := mat((0,1,0,0),(1,0,0,0),(0,0,0,1),(0,0,1,0))$
r6 := mat((0,1,1,0),(1,0,0,1),(1,0,0,1),(0,1,1,0))$
r7 := mat((j,b,b,0),(b, - j,0,b),(b,0, - j,b),(0,b,b,j))$

MATEIGEN(R7,eta);


{{ - 4*b**2 - j**2 + eta**2,1,
mat((arbcomplex(1)),((2*arbcomplex(1)*b)/(j + eta)),
((2*arbcomplex(1)*b)/(j + eta)),(arbcomplex(1)))$},
{ - j + eta,1,
mat(( - arbcomplex(2)),(0),(0),(arbcomplex(2)))$},
{j + eta,1,
mat((0),( - arbcomplex(3)),(arbcomplex(3)),(0))$}}$
```

## 27. Two-Point Heisenberg Model

We consider the two-point Heisenberg model

$$\hat{H} = J \sum_{j=1}^{2} \mathbf{S}_j \cdot \mathbf{S}_{j+1}, \tag{1}$$

where $J$ is the so-called exchange constant ($J > 0$ or $J < 0$) and $\cdot$ denotes the scalar product. We impose cyclic boundary conditions, i.e. $\mathbf{S}_3 = \mathbf{S}_1$. It follows that

$$\hat{H} = J(\mathbf{S}_1 \cdot \mathbf{S}_2 + \mathbf{S}_2 \cdot \mathbf{S}_3) \equiv J(\mathbf{S}_1 \cdot \mathbf{S}_2 + \mathbf{S}_2 \cdot \mathbf{S}_1). \tag{2}$$

From (2) it follows that

$$\hat{H} = J(S_{x,1} S_{x,2} + S_{y,1} S_{y,2} + S_{z,1} S_{z,2} + S_{x,2} S_{x,1} + S_{y,2} S_{y,1} + S_{z,2} S_{z,1}). \tag{3}$$

Since

$$S_{x,1} := S_x \otimes I, \qquad S_{x,2} := I \otimes S_x \tag{4}$$

etc., where $I$ is the $2 \times 2$ unit matrix and $\otimes$ is the Kronecker product (see 26), it follows that

$$\hat{H} = J((S_x \otimes I)(I \otimes S_x) + (S_y \otimes I)(I \otimes S_y) + (S_z \otimes I)(I \otimes S_z) +$$

$$(I \otimes S_x)(S_x \otimes I) + (I \otimes S_y)(S_y \otimes I) + (I \otimes S_z)(S_z \otimes I)). \tag{5}$$

Therefore

$$\hat{H} = 2J((S_x \otimes S_x) + (S_y \otimes S_y) + (S_z \otimes S_z)). \tag{6}$$

Since the spin matrices are given by

$$S_x := \frac{1}{2} \begin{pmatrix} 0 & 1 \\ 1 & 0 \end{pmatrix}, \qquad S_y := \frac{1}{2} \begin{pmatrix} 0 & -i \\ i & 0 \end{pmatrix}, \qquad S_z := \frac{1}{2} \begin{pmatrix} 1 & 0 \\ 0 & -1 \end{pmatrix} \tag{7}$$

we obtain

$$S_x \otimes S_x = \frac{1}{4} \begin{pmatrix} 0 & 1 \\ 1 & 0 \end{pmatrix} \otimes \begin{pmatrix} 0 & 1 \\ 1 & 0 \end{pmatrix} = \frac{1}{4} \begin{pmatrix} 0 & 0 & 0 & 1 \\ 0 & 0 & 1 & 0 \\ 0 & 1 & 0 & 0 \\ 1 & 0 & 0 & 0 \end{pmatrix} \tag{8a}$$

$$S_y \otimes S_y = \frac{1}{4} \begin{pmatrix} 0 & -i \\ i & 0 \end{pmatrix} \otimes \begin{pmatrix} 0 & -i \\ i & 0 \end{pmatrix} = \frac{1}{4} \begin{pmatrix} 0 & 0 & 0 & -1 \\ 0 & 0 & 1 & 0 \\ 0 & 1 & 0 & 0 \\ -1 & 0 & 0 & 0 \end{pmatrix} \qquad (8b)$$

$$S_z \otimes S_z = \frac{1}{4} \begin{pmatrix} 1 & 0 \\ 0 & -1 \end{pmatrix} \otimes \begin{pmatrix} 1 & 0 \\ 0 & -1 \end{pmatrix} = \frac{1}{4} \begin{pmatrix} 1 & 0 & 0 & 0 \\ 0 & -1 & 0 & 0 \\ 0 & 0 & -1 & 0 \\ 0 & 0 & 0 & 1 \end{pmatrix}. \qquad (8c)$$

Then the Hamilton operator $\hat{H}$ is given by the $4 \times 4$ symmetric matrix

$$\hat{H} = \frac{J}{2} \begin{pmatrix} 1 & 0 & 0 & 0 \\ 0 & -1 & 2 & 0 \\ 0 & 2 & -1 & 0 \\ 0 & 0 & 0 & 1 \end{pmatrix} \equiv \frac{J}{2} \left( (1) \oplus \begin{pmatrix} -1 & 2 \\ 2 & -1 \end{pmatrix} \oplus (1) \right), \qquad (9)$$

where $\oplus$ denotes the *direct sum*.

In the program we evaluate the eigenvalues and eigenvectors of $\hat{H}$. The eigenvalues are given by

$$E_1 = \frac{J}{2} \quad \text{three times degenerate}, \qquad E_2 = \frac{-3J}{2}. \qquad (10)$$

If $J$ is positive, then $E_2$ is the ground state energy. If $J$ is negative, then $E_1$ is the ground state energy.

```
%heismod.red;

procedure Kron(A,B);
begin
n := 2; m := n*n;

operator A$
matrix AA(n,n)$
for i:=1:n do
for j:=1:n do
AA(i,j):=A(i,j)$

operator B$
matrix BB(n,n)$
for i:=1:n do
for j:=1:n do
BB(i,j):=B(i,j)$

operator C$
matrix CC(m,m);
c1 := 0; c2 := 0;
for r:=1:n do
for s:=1:n do
begin
for i:=1:n do
for j:=1:n do
begin
c1 := n*(r-1); c2 := n*(s-1);
CC(i+c1,j+c2) := AA(r,s)*BB(i,j);
end;
end;
return CC;
end;     % end procedure Kron;

SX(1,1) := 0; SX(1,2) := 1/2; SX(2,1) := 1/2; SX(2,2) := 0;
SY(1,1) := 0; SY(1,2) := -i/2; SY(2,1) := i/2; SY(2,2) := 0;
SZ(1,1) := 1/2; SZ(1,2) := 0; SZ(2,1) := 0; SZ(2,2) := -1/2;

R1 := Kron(SX,SX);
R2 := Kron(SY,SY);
R3 := Kron(SZ,SZ);

R := 2*J*(R1 + R2 + R3);
```

```
mateigen(R,eta);
```

The output is

```
R1 := MAT((0,0,0,1/4),(0,0,1/4,0),(0,1/4,0,0),(1/4,0,0,0))$

R2 := MAT((0,0,0,(-1)/4),(0,0,1/4,0),(0,1/4,0,0),((-1)/4,0,0,0))$

R3 := MAT((1/4,0,0,0),(0,(-1)/4,0,0),(0,0,(-1)/4,0),(0,0,0,1/4))$

R := MAT((J/2,0,0,0),(0,(-J)/2,J,0),(0,J,(-J)/2,0),(0,0,0,J/2))$

{{ - J + 2*ETA,3,
MAT((ARBCOMPLEX(1)),(ARBCOMPLEX(2)),(ARBCOMPLEX(2)),
(ARBCOMPLEX(3)))$},
{3*J + 2*ETA,1,
MAT((0),( - ARBCOMPLEX(4)),(ARBCOMPLEX(4)),(0))$}}$
```

*Remark:* This list gives the eigenvalues $\eta = J$ (3 times) and $\eta = -3J/2$ with the corresponding eigenvectors.

## 28. Fermi Operators

Consider a family of linear operators

$$\{ c_j, c_j^\dagger, \qquad j = 1, 2, \ldots, n \} \tag{1}$$

defined on a finite dimensional vector space $V$ satisfying the *anticommutation relations*

$$[c_j, c_k]_+ = [c_j^\dagger, c_k^\dagger]_+ = 0 \tag{2}$$

$$[c_j, c_k^\dagger]_+ = \delta_{jk} I, \tag{3}$$

where 0 is the zero operator, $I$ the unit operator and $j, k = 1, 2, \ldots, n$. Notice that $[X, Y]_+ \equiv XY + YX$. Operators satisfying (2) and (3) are called annihilation and creation operators for fermions. In the present case we have spinless Fermi operators. In problem 29 we introduce Fermi operators with spin. The indices are the quantum numbers.

We define a state $|0\rangle$ (the so-called *vacuum state*) with the properties

$$c_j|0\rangle = 0, \qquad j = 1, 2, \ldots, n \tag{4}$$

$$\langle 0|0\rangle = 1. \tag{5}$$

Thus we assume that $|0\rangle$ is normalized. Other states can now be constructed from $|0\rangle$ and the creation operators. From (4) it follows that

$$\langle 0|c_j^\dagger = 0, \qquad j = 1, 2, \ldots, n. \tag{6}$$

From (2) it follows that

$$c_j^2 = 0, \qquad \text{and} \qquad c_j^{\dagger 2} = 0. \tag{7}$$

In the program we denote the Fermi creation operator $c_j^\dagger$ by $d(j)$ and the Fermi annihilation operator $c_j$ by $c(j)$. The vacuum state $|0\rangle$ is denoted by $vs(0)$ and the dual vacuum state $\langle 0|$ is denoted by $ds(0)$.

*Example 1:* Applying the rules given above we find that

$$c_1 c_2 c_1^\dagger c_2^\dagger c_3^\dagger = -c_3^\dagger - c_2^\dagger c_3^\dagger c_2 - c_1^\dagger c_3^\dagger c_1 + c_1^\dagger c_2^\dagger c_3^\dagger c_1 c_2. \tag{8}$$

*Example 2:* Applying the rules given above we find that

$$(c_1^\dagger c_1 + c_2^\dagger c_2)^2 = c_2^\dagger c_2 - c_2^\dagger c_1^\dagger c_2 c_1 + c_1^\dagger c_1 - c_1^\dagger c_2^\dagger c_1 c_2. \tag{9}$$

*Example 3:* Consider the Hamilton operator

$$\hat{H} = B(c_1^\dagger c_2 + c_2^\dagger c_1 + c_2^\dagger c_3 + c_3^\dagger c_2) + U(c_1^\dagger c_1 c_2^\dagger c_2 + c_2^\dagger c_2 c_3^\dagger c_3). \tag{10}$$

The total number operator is given by

$$\hat{N} := c_1^\dagger c_1 + c_2^\dagger c_2 + c_3^\dagger c_3. \tag{11}$$

We evaluate the commutator $[\hat{H}, \hat{N}]$ and obviously find that $\hat{H}$ and $\hat{N}$ commute, i.e.

$$[\hat{H}, \hat{N}] = 0. \tag{12}$$

Then we evaluate the matrix representation of $\hat{H}$, where we consider the case with two Fermi particles. The basis is given by

$$\{ c_1^\dagger c_2^\dagger |0\rangle, \quad c_1^\dagger c_3^\dagger |0\rangle, \quad c_2^\dagger c_3^\dagger |0\rangle \}. \tag{13}$$

The dual basis takes the form

$$\{ \langle 0|c_2 c_1, \quad \langle 0|c_3 c_1, \quad \langle 0|c_3 c_2 \}. \tag{14}$$

The matrix elements of $\hat{H}$ are given by

$$H_{jk} := \langle \phi_j|\hat{H}|\phi_k\rangle. \tag{15}$$

We find the symmetric matrix

$$\begin{pmatrix} U & B & 0 \\ B & 0 & B \\ 0 & B & U \end{pmatrix}. \tag{16}$$

```
%fermi.red;

%The operators c(j) are the annihilation operators;
%The operators d(j) are the creation operators;

operator c, d, vs, ds;
noncom c, d, vs, ds;

for all j let c(j)**2=0;
for all j let d(j)**2=0;
for all j let c(j)*d(j)=-d(j)*c(j)+1;
for all j,k such that j neq k let c(j)*d(k)=-d(k)*c(j);
for all j,k such that j leq k let c(j)*c(k)=-c(k)*c(j);
for all j,k such that j leq k let d(j)*d(k)=-d(k)*d(j);
for all j,k let c(j)*vs(k)=0;   % equation (3);
for all j,k let ds(k)*d(j)=0;   % equation (5);
let ds(0)*vs(0)=1;

%Example 1;
result1 := c(1)*c(2)*d(1)*d(2)*d(3);

%Example 2;
result2 := (d(1)*c(1)+d(2)*c(2))**2;

%Example 3;
N := d(1)*c(1) + d(2)*c(2) + d(3)*c(3);
H := B*(d(1)*c(2) + d(2)*c(1) +  d(2)*c(3) + d(3)*c(2)) +
U*(d(1)*c(1)*d(2)*c(2) + d(2)*c(2)*d(3)*c(3));
comm := H*N - N*H;
matrix A(3,3);
A(1,1) := ds(0)*c(2)*c(1)*H*d(1)*d(2)*vs(0);
A(1,2) := ds(0)*c(2)*c(1)*H*d(1)*d(3)*vs(0);
A(1,3) := ds(0)*c(2)*c(1)*H*d(2)*d(3)*vs(0);
A(2,1) := ds(0)*c(3)*c(1)*H*d(1)*d(2)*vs(0);
A(2,2) := ds(0)*c(3)*c(1)*H*d(1)*d(3)*vs(0);
A(2,3) := ds(0)*c(3)*c(1)*H*d(2)*d(3)*vs(0);
A(3,1) := ds(0)*c(3)*c(2)*H*d(1)*d(2)*vs(0);
A(3,2) := ds(0)*c(3)*c(2)*H*d(1)*d(3)*vs(0);
A(3,3) := ds(0)*c(3)*c(2)*H*d(2)*d(3)*vs(0);
```

The output is

```
RESULT1:=D(3)*D(2)*C(2)+D(3)*D(2)*D(1)*C(2)*C(1)+D(3)*D(1)*C(1)-D(3)$

RESULT2:=D(2)*C(2)-2*D(2)*D(1)*C(2)*C(1)+D(1)*C(1)$

COMM := 0$

A(1,1) := U$
A(1,2) := B$
A(1,3) := 0$
A(2,1) := B$
A(2,2) := 0$
A(2,3) := B$
A(3,1) := 0$
A(3,2) := B$
A(3,3) := U$
```

## 29. Fermi Operators with Spin and the Hubbard Model

Here we study Fermi operators with spin and give an application to the Hubbard model. Consider a family of linear operators $c_{j\sigma}$, $c_{j\sigma}^\dagger$, $j = 1, 2, \ldots, n$ defined on a finite dimensional vector space $V$ satisfying the anticommutation relations

$$[c_{j\sigma}, c_{k\sigma'}]_+ = [c_{j\sigma}^\dagger, c_{k\sigma'}^\dagger]_+ = 0, \qquad [c_{j\sigma}, c_{k\sigma'}^\dagger]_+ = \delta_{jk}\delta_{\sigma\sigma'}I, \tag{1}$$

where 0 is the zero operator, $I$ the unit operator, $j, k = 1, 2, \ldots n$ and

$$\sigma, \sigma' \in \{\uparrow, \downarrow\}.$$

Operators satisfying (1) are called annihilation and creation operators for fermions with spin. We define a state $|0\rangle$ (the so-called vacuum state) with the property

$$c_{j\sigma}|0\rangle = 0 \qquad j = 1, 2, \ldots n, \qquad \sigma \in \{\uparrow, \downarrow\}, \tag{2}$$

where

$$\langle 0|0\rangle = 1. \tag{3}$$

Thus we assume that $|0\rangle$ is normalized. Other states can now be constructed from $|0\rangle$ and the creation operators. For example $c_{1\uparrow}^\dagger c_{4\downarrow}^\dagger|0\rangle$. The indexes $j$, $k$ are the quantum numbers together with the spin $\sigma$. From the relations given above it follows that

$$(c_{j\sigma}^\dagger)^2 = 0, \qquad (c_{j\sigma})^2 = 0, \qquad j = 1, 2, \ldots, n. \tag{4}$$

We also have to introduce an ordering for the spin. We put all spin up operators on the left hand side. Furthermore we have to introduce an ordering for the quantum number (index) $j$, where $j = 1, 2, \ldots, n$. We set the Fermi operators with the lower quantum number on the left hand side, i.e. $j_1 \leq j_2 \leq \ldots \leq j_n$.

As an application we consider the four point Hubbard model with cyclic boundary conditions. We give a higher order conserved quantity. The Hubbard model plays an important role in the modeling of magnetism, charge density waves and high-$T_c$ superconductivity, since the interaction term of the Hubbard Hamiltonian can be written as

$$n_{i\uparrow}n_{i\downarrow} \equiv \frac{1}{4}(1 - \alpha_i) + \hat{R}_{iz} + \frac{1}{3}(\alpha_i - 1)S_i^2 + \frac{1}{3}(\alpha_i + 1)\boldsymbol{R}_i^2 \tag{5}$$

where $n_{j\sigma} := c_{j\sigma}^\dagger c_{j\sigma}$. Here $\boldsymbol{S}_i$ are the spin operators

$$\hat{S}_{ix} = \frac{1}{2}(c_{i\uparrow}^\dagger c_{i\downarrow} + c_{i\downarrow}^\dagger c_{i\uparrow}), \qquad \hat{S}_{iy} = \frac{1}{2i}(c_{i\uparrow}^\dagger c_{i\downarrow} - c_{i\downarrow}^\dagger c_{i\uparrow}), \qquad \hat{S}_{iz} = \frac{1}{2}(n_{i\uparrow} - n_{i\downarrow}) \tag{6}$$

and $R_i$ are the quasi-spin operators

$$\hat{R}_{ix} = \frac{1}{2}(c_{i\uparrow}^{\dagger}c_{i\downarrow}^{\dagger} + c_{i\downarrow}c_{i\uparrow}), \qquad \hat{R}_{iy} = \frac{1}{2i}(c_{i\uparrow}^{\dagger}c_{i\downarrow}^{\dagger} - c_{i\uparrow}c_{i\downarrow}), \qquad \hat{R}_{iz} = \frac{1}{2}(n_{i\uparrow} + n_{i\downarrow} - 1). \quad (7)$$

Both the spin operators and quasi-spin operators form a Lie algebra under the commutator. This decomposition of the interacting part makes it possible to search for magnetism ($\langle S_i \rangle \neq 0$), charge ordering ($\langle R_{iz} \rangle \neq 0$) or superconductivity ($\langle R_{ix} \rangle \neq 0$) in the Hubbard model.

The Hubbard model commutes with $\hat{N}_e$ and $\hat{S}_z$ and therefore the spectrum can be calculated in each of the subspaces separately. The study is limited to the half-filled case, i.e. $N_e = N$ (where $N$ is the number of lattice sites and $N_e$ is the number of electrons) and with total spin in the $z$-direction $S_z = 0$. In Wannier representation the Hubbard model is given by

$$\hat{H} = t\sum_{i=1}^{4} \sum_{\sigma \in \{\uparrow\downarrow\}} (c_{i+1\sigma}^{\dagger}c_{i\sigma} + c_{i\sigma}^{\dagger}c_{i+1\sigma}) + U\sum_{i=1}^{4} n_{i\uparrow}n_{i\downarrow}, \qquad (8)$$

where $5 \equiv 1$. Thus the hopping integral $t$ only acts for nearest neighbours. The total number operator is given by

$$\hat{N}_e := \sum_{i=1}^{4} \sum_{\sigma \in \{\uparrow\downarrow\}} n_{i\sigma} \qquad (9)$$

with the eigenvalues $N_e = 0, 1, 2, 3, 4$. The total spin operator in $z$ direction is given by

$$\hat{S}_z := \frac{1}{2}\sum_{i=1}^{4}(n_{i\uparrow} - n_{i\downarrow}) \qquad (10)$$

with the eigenvalues $0, 1/2, -1/2, 1, -1, 3/2, -3/2, 2, -2$. For $N_e = 4$, $S_z = 0$ the dimension of the Hilbert space is given by $\dim \mathcal{H} = 36$. A basis is given by

$$\{c_{i\uparrow}^{\dagger}c_{j\uparrow}^{\dagger}c_{m\downarrow}^{\dagger}c_{n\downarrow}^{\dagger}|0\rangle; \quad i < j, \ m < n; \ i = 1, 2, 3; \ m = 1, 2, 3\}. \qquad (11)$$

```
%hubbard.red;

%c1(j): fermi creation operator with spin up;
%c2(j): fermi annihilation operator with spin up;
%d1(j): fermi creation operator with spin down;
%d2(j): fermi annihilation operator with spin down;

operator c1, d1, c2, d2, N, HK, HU, CL;
noncom c1, d1, c2, d2, N, HK, HU, CL;

for all j let c1(j)*c1(j) = 0;
for all j let c2(j)*c2(j) = 0;
for all j let d1(j)*d1(j) = 0;
for all j let d2(j)*d2(j) = 0;

for all j let c2(j)*c1(j) = - c1(j)*c2(j) + 1;
for all j,k such that j neq k let
c2(j)*c1(k) = - c1(k)*c2(j);

for all j let d2(j)*d1(j) = - d1(j)*d2(j) + 1;
for all j,k such that j neq k let
d2(j)*d1(k) = - d1(k)*d2(j);

for all j,k let d1(j)*c1(k) = - c1(k)*d1(j);
for all j,k let d2(j)*c2(k) = - c2(k)*d2(j);
for all j,k let d1(j)*c2(k) = - c2(k)*d1(j);
for all j,k let d2(j)*c1(k) = - c1(k)*d2(j);

for all j,k such that j leq k let c1(j)*c1(k) = - c1(k)*c1(j);
for all j,k such that j leq k let c2(j)*c2(k) = - c2(k)*c2(j);
for all j,k such that j leq k let d1(j)*d1(k) = - d1(k)*d1(j);
for all j,k such that j leq k let d2(j)*d2(k) = - d2(k)*d2(j);

S1X := (1/2)*(c1(1)*d2(1)+d1(1)*c2(1));
S1Y := (-i/2)*(c1(1)*d2(1)-d1(1)*c2(1));


r1 := S1X*S1Y - S1Y*S1X;

HK := t*(c1(2)*c2(1)+c1(3)*c2(2)+c1(4)*c2(3)+c1(1)*c2(4)
    +c1(1)*c2(2)+c1(2)*c2(3)+c1(3)*c2(4)+c1(4)*c2(1)
    +d1(2)*d2(1)+d1(3)*d2(2)+d1(4)*d2(3)+d1(1)*d2(4)
    +d1(1)*d2(2)+d1(2)*d2(3)+d1(3)*d2(4)+d1(4)*d2(1));
HU := U*(c1(1)*c2(1)*d1(1)*d2(1)+c1(2)*c2(2)*d1(2)*d2(2)
    +c1(3)*c2(3)*d1(3)*d2(3)+c1(4)*c2(4)*d1(4)*d2(4));
```

```
NU := c1(1)*c2(1)+c1(2)*c2(2)+c1(3)*c2(3)+c1(4)*c2(4);

ND := d1(1)*d2(1)+d1(2)*d2(2)+d1(3)*d2(3)+d1(4)*d2(4);

r2 := HK*NU - NU*HK;
r3 := HK*ND - ND*HK;
r4 := HU*NU - NU*HU;
r5 := HU*ND - ND*HU;

CL := (c1(1)*c2(4)-c1(4)*c2(1))*(d1(1)*d2(1)+d1(4)*d2(4))
    +(d1(1)*d2(4)-d1(4)*d2(1))*(c1(1)*c2(1)+c1(4)*c2(4))
    -(c1(1)*c2(4)-c1(4)*c2(1)+d1(1)*d2(4)-d1(4)*d2(1))
    +(c1(2)*c2(1)-c1(1)*c2(2))*(d1(2)*d2(2)+d1(1)*d2(1))
    +(d1(2)*d2(1)-d1(1)*d2(2))*(c1(2)*c2(2)+c1(1)*c2(1))
    -(c1(2)*c2(1)-c1(1)*c2(2)+d1(2)*d2(1)-d1(1)*d2(2))
    +(c1(3)*c2(2)-c1(2)*c2(3))*(d1(3)*d2(3)+d1(2)*d2(2))
    +(d1(3)*d2(2)-d1(2)*d2(3))*(c1(3)*c2(3)+c1(2)*c2(2))
    -(c1(3)*c2(2)-c1(2)*c2(3)+d1(3)*d2(2)-d1(2)*d2(3))
    +(c1(4)*c2(3)-c1(3)*c2(4))*(d1(4)*d2(4)+d1(3)*d2(3))
    +(d1(4)*d2(3)-d1(3)*d2(4))*(c1(4)*c2(4)+c1(3)*c2(3))
    -(c1(4)*c2(3)-c1(3)*c2(4)+d1(4)*d2(3)-d1(3)*d2(4));

r6 := HK*CL - CL*HK;

r7 := HU*CL - CL*HU;
```

The output is given by

```
R1 := (I*(C1(1)*C2(1) - D1(1)*D2(1)))/2$

R2 := 0$

R3 := 0$

R4 := 0$

R5 := 0$

R6 := 0$

R7 := 0$
```

## 30. Bose Operators

Consider the linear operators $b$ and $b^\dagger$ defined on an inner product space. The commutation relations for the so-called Bose operators $b$ and $b^\dagger$ are given by

$$[b, b^\dagger] = I, \tag{1}$$

where $I$ is the identity operator and

$$[b, b] = [b^\dagger, b^\dagger] = 0. \tag{2}$$

Here 0 is the zero operator and $[\,,\,]$ denotes the commutator. From (1) it follows that

$$bb^\dagger = I + b^\dagger b. \tag{3}$$

This equation is implemented in the program.

Let $|0\rangle$ be the vacuum state, i.e.

$$b|0\rangle = 0, \qquad \langle 0|0\rangle = 1. \tag{4}$$

From (4) it follows that

$$0 = \langle 0|b^\dagger. \tag{5}$$

A state is given by

$$(b^\dagger)^n|0\rangle, \tag{6}$$

where $n = 0, 1, 2, \ldots$ . However, this state is not normalized. Normalizing this state yields

$$\frac{1}{\sqrt{n!}}(b^\dagger)|0\rangle. \tag{7}$$

In the program all the annihilation operators $b$ are shifted to the right. Then property (4), i.e. $b|0\rangle = 0$, can be applied if the operator acts on the vacuum state $|0\rangle$.

*Remark:* In the program $b(2)$ denotes the Bose creation operator and $b(1)$ denotes the Bose annihilation operator.

```
%bose.red;

operator b;
noncom b;
%b(1) denotes the Bose annihilation operator;
%b(2) denotes the Bose creation operator;
%The next line defines the commutation rules;
%All the annihilation operators are shifted to the right;

let b(1)*b(2) = b(2)*b(1) + 1;

%Example 1;
result1 := b(1)*b(2)*b(1)*b(2);
%Example 2;
result2 := b(1)*b(2)*b(2);
%Example 3;
result3 := (b(1)+b(2))**4;
%Example 4;
result4 := b(1) + b(1)*b(2) + b(1)*b(2)*b(2) + b(1)*b(2)*b(2)*b(2);

%Let us assume that the annihilation operator b(1)
%acts on the vacuum state;

let b(1) = 0;
result1; result2; result3; result4;
```

The output is

```
result1 := b(2)**2*b(1)**2 + 3*b(2)*b(1) + 1$
result2 := b(2)**2*b(1) + 2*b(2)$
result3 := b(2)**4 + 4*b(2)**3*b(1) + 6*b(2)**2*b(1)**2 + 6*b
(2)**2 + 4*b(2)*b(1)**3 + 12*b(2)*b(1) + b(1)**4 + 6*b(1)**2 + 3$
result4 := b(2)**3*b(1) + b(2)**2*b(1) + 3*b(2)**2 + b(2)*b(1) +
2*b(2) + b(1) + 1$


1$
2*b(2)$
b(2)**4 + 6*b(2)**2 + 3$
3*b(2)**2 + 2*b(2) + 1$
```

## 31. Matrix Representation of Bose Operators

We again consider the Bose operators $\{\, b,\ b^\dagger \,\}$ and the normalized states

$$|n\rangle := \frac{1}{\sqrt{n!}}(b^\dagger)^n|0\rangle, \tag{1}$$

where $n = 0, 1, 2, \ldots$. The dual state is given by

$$\langle n| = \langle 0|b^n \frac{1}{\sqrt{n!}}, \tag{2}$$

where $n = 1, 2, \ldots$.

In the program we introduce the rules

$$b|n\rangle = \sqrt{n}|n-1\rangle \tag{3a}$$

$$b^\dagger|n\rangle = \sqrt{n+1}|n+1\rangle \tag{3b}$$

$$\langle n|b = \langle n+1|\sqrt{n+1} \tag{3c}$$

$$\langle n|b^\dagger = \langle n-1|\sqrt{n}. \tag{3d}$$

Furthermore we introduce the orthogonality relation

$$\langle n|m\rangle = \delta_{n,m}. \tag{4}$$

In the program $b$ denotes the annihilation operator and $bd$ denotes the creation operator.

For example 1 we evaluate $bb|n\rangle$, in example 2 we determine $\langle 2|3\rangle$ and $\langle 4|4\rangle$, while in example 3 we calculate $\langle n|b|n\rangle$, $\langle n|b|n+1\rangle$ and $\langle n+1|b|n\rangle$.

```
%matbos.red;

operator b, bd, no, mo;
noncom b, bd, no, mo;

%no and mo are states, mo is the dual state of no;
for all n let b*no(n) = sqrt(n)*no(n-1);
for all n let bd*no(n) = sqrt(n+1)*no(n+1);
for all n let mo(n)*b = mo(n+1)*sqrt(n+1);
for all n let mo(n)*bd = mo(n-1)*sqrt(n);
for all n let mo(n)*no(n) = 1;
for all j, k such that j neq k let mo(j)*no(k) = 0;

%Example 1;
r1 := b*(b*no(n));
n := 1;
r1;
clear n;

%Example 2;
mo(2)*no(3);
mo(4)*no(4);

%Example 3;
r2 := b*no(n);
r3 := mo(n)*r2;
r4 := b*no(n+1);
r5 := mo(n)*r4;
r6 := b*no(n);
r7 := mo(n+1)*r6;
```

The output is

```
R1 := SQRT(N)*SQRT(N - 1)*NO(N - 2)$
0$

0$
1$

R2 := SQRT(N)*NO(N - 1)$   R3 := 0$
R4 := SQRT(N + 1)*NO(N)$   R5 := SQRT(N + 1)$
R6 := SQRT(N)*NO(N - 1)$   R7 := 0$
```

## 32. Coherent States

Let $b$ be a Bose annihilation operator. Consider the eigenvalue problem

$$b|\beta\rangle = \beta|\beta\rangle. \tag{1}$$

The eigenstate $|\beta\rangle$ is called a *Bose coherent state*. From (1) it follows that

$$\langle\beta|b^\dagger = \langle\beta|\beta^*. \tag{2}$$

With the help of the vacuum state $|0\rangle$ the coherent state can be written as

$$|\beta\rangle = \exp(-|\beta|^2/2) \sum_{n=0}^{\infty} \frac{\beta^n}{\sqrt{n!}}|n\rangle \equiv \exp(-|\beta|^2/2) \exp(\beta b^\dagger|0\rangle), \tag{3}$$

where $\beta \in C$ and $C$ denotes the complex numbers. The coherent states are not orthogonal, but we have

$$\langle\beta|\gamma\rangle = \exp(-(|\gamma|^2 + |\beta|^2)/2 + \gamma\bar{\beta}), \tag{4}$$

where

$$b|\gamma\rangle = \gamma|\gamma\rangle. \tag{5}$$

The completeness relation is given by

$$\int_C |\beta\rangle\langle\beta|\frac{d^2\beta}{\pi} = I, \tag{6}$$

where $I$ is the identity operator and the integration is over the entire complex plane $C$.

In the program we implement (3) and (4), where $bd$ denotes $b^\dagger$, $cs(z)$ denotes the coherent state $|\beta\rangle$ and $ds(z)$ denotes $\langle\beta|$.

```
%coher.red;

operator b, bd, cs, ds;
noncom b, bd, cs, ds;

for all z let b*cs(z) = z*cs(z);
for all z let ds(z)*bd = ds(z)*z;
for all z, w let ds(z)*cs(w) =
exp(-(1/2)*(z*conj(z)+w*conj(w)-2*conj(z)*w));

%Example 1;
r1 := b*(b*cs(z));
z := 1;
r1;
clear z;

%Example 2;
ds(z)*cs(z);

%Example 3;
r2 := b*cs(z);
r3 := ds(w)*r2;
```

The output is

```
R1 := Z**2*CS(Z)$

CS(1)$

1$
R2 := Z*CS(Z)$
R3 := (E**((IMPART(W)*I*W + IMPART(Z)*I*Z + 2*REPART(W)*Z)/2)*Z)/
E**((2*IMPART(W)*I*Z + REPART(W)*W + REPART(Z)*Z)/2)$
```

## 33. Quartic Hamilton Operator and Bose Operators

Consider a family of linear operators $b_j$, $b_j^\dagger$, $1 \leq j \leq m$, on an inner product space $V$, satisfying the commutation relation

$$[b_j, b_k] = [b_j^\dagger, b_k^\dagger] = 0, \qquad [b_j, b_k^\dagger] = \delta_{jk} I, \tag{1}$$

where $I$ is the identity operator. The operator $b_j^\dagger$ is called a creation operator for bosons and its adjoint $b_j$ is called an annihilation operator for bosons.

We consider the Hamilton operator

$$\hat{H} = \frac{1}{2}(\hat{p}_1^2 + \hat{p}_2^2 + x_1^2 + x_2^2) + a x_1^2 x_2^2, \tag{2}$$

where $a$ is a positive constant and (we set $\hbar = 1$)

$$\hat{p}_j = -i \frac{\partial}{\partial x_j}. \tag{3}$$

We express the Hamilton operator $\hat{H}$ with the help of Bose operators, where

$$x_1 := \frac{1}{\sqrt{2}}(b_1^\dagger + b_1) \tag{4a}$$

$$x_2 := \frac{1}{\sqrt{2}}(b_2^\dagger + b_2) \tag{4b}$$

$$\hat{p}_1 := \frac{i}{\sqrt{2}}(b_1^\dagger - b_1) \tag{4c}$$

$$\hat{p}_2 := \frac{i}{\sqrt{2}}(b_2^\dagger - b_2) \tag{4d}$$

```
%quartic.red;

%bd(j) Bose creation operator with quantum number j;
%b(j) Bose annihilation operator with quantum number j;

operator x, p, b, bd;
noncom x, p, b, bd;

for all j let b(j)*bd(j) = bd(j)*b(j) + 1;
for all j,k such that j neq k let
b(j)*bd(k) = bd(k)*b(j);

for all j,k such that j leq k let b(j)*b(k) = b(k)*b(j);
for all j,k such that j leq k let bd(j)*bd(k) = bd(k)*bd(j);

x(1) := (bd(1) + b(1))/sqrt(2);
x(2) := (bd(2) + b(2))/sqrt(2);
p(1) := i*(bd(1) - b(1))/sqrt(2);
p(2) := i*(bd(2) - b(2))/sqrt(2);

HO := p(1)*p(1)/2 + p(2)*p(2)/2 + x(1)*x(1)/2 + x(2)*x(2)/2;
H1 := a*x(1)*x(1)*x(2)*x(2);

HO + H1;
```

The output is

```
HO := BD(2)*B(2) + BD(1)*B(1) + 1$

H1 := (A*(B(2)**2*B(1)**2 + B(2)**2 + B(1)**2 + BD(2)**2*B(1)**2 +
BD(2)**2*BD(1)**2 + 2*BD(2)**2*BD(1)*B(1) + BD(2)**2 +
2*BD(2)*B(2)*B(1)**2 + 2*BD(2)*B(2) + 2*BD(2)*BD(1)**2*B(2) +
4*BD(2)*BD(1)*B(2)*B(1) + BD(1)**2*B(2)**2 + BD(1)**2 +
2*BD(1)*B(2)**2*B(1) + 2*BD(1)*B(1) + 1))/4$

HO + H1;

(B(2)**2*B(1)**2*A + B(2)**2*A + B(1)**2*A + BD(2)**2*B(1)**2*A +
BD(2)**2*BD(1)**2*A + 2*BD(2)**2*BD(1)*B(1)*A + BD(2)**2*A +
2*BD(2)*B(2)*B(1)**2*A + 2*BD(2)*B(2)*A + 4*BD(2)*B(2) +
2*BD(2)*BD(1)**2*B(2)*A + 4*BD(2)*BD(1)*B(2)*B(1)*A +
BD(1)**2*B(2)**2*A + BD(1)**2*A + 2*BD(1)*B(2)**2*B(1)*A +
2*BD(1)*B(1)*A + 4*BD(1)*B(1) + A + 4)/4$
```

## 34. Dirac Equation and Dispersion Law

The *Dirac equation* for the free particle is given by

$$-i\hbar \sum_{j=1}^{4} \frac{\partial}{\partial x_j} \gamma_j \Psi = im_0 c \Psi, \tag{1}$$

where $x_4 = ct$ and

$$\Psi = \begin{pmatrix} u_1 \\ u_2 \\ u_3 \\ u_4 \end{pmatrix}. \tag{2}$$

Here $m_0$ is the mass of the particle and the $\gamma_j$ are the $\gamma$ matrices. The gamma matrices are given by

$$\gamma_1 := \begin{pmatrix} 0 & 0 & 0 & -i \\ 0 & 0 & -i & 0 \\ 0 & i & 0 & 0 \\ i & 0 & 0 & 0 \end{pmatrix}, \qquad \gamma_2 := \begin{pmatrix} 0 & 0 & 0 & -1 \\ 0 & 0 & 1 & 0 \\ 0 & 1 & 0 & 0 \\ -1 & 0 & 0 & 0 \end{pmatrix}$$

$$\gamma_3 := \begin{pmatrix} 0 & 0 & -i & 0 \\ 0 & 0 & 0 & i \\ i & 0 & 0 & 0 \\ 0 & -i & 0 & 0 \end{pmatrix}, \qquad \gamma_4 := \begin{pmatrix} 1 & 0 & 0 & 0 \\ 0 & 1 & 0 & 0 \\ 0 & 0 & -1 & 0 \\ 0 & 0 & 0 & -1 \end{pmatrix}. \tag{3}$$

From (1) we obtain

$$\begin{pmatrix} -i\partial u_4/\partial x_1 - \partial u_4/\partial x_2 - i\partial u_3/\partial x_3 + \partial u_1/\partial x_4 + m_0 c u_1/\hbar \\ -i\partial u_3/\partial x_1 + \partial u_3/\partial x_2 + i\partial u_4/\partial x_3 + \partial u_2/\partial x_4 + m_0 c u_2/\hbar \\ i\partial u_2/\partial x_1 + \partial u_2/\partial x_2 + i\partial u_1/\partial x_3 - \partial u_3/\partial x_4 + m_0 c u_3/\hbar \\ i\partial u_1/\partial x_1 - \partial u_1/\partial x_2 - i\partial u_2/\partial x_3 - \partial u_4/\partial x_4 + m_0 c u_4/\hbar \end{pmatrix} = \begin{pmatrix} 0 \\ 0 \\ 0 \\ 0 \end{pmatrix}. \tag{4}$$

Let

$$\mathbf{k} \cdot \mathbf{x} = k_1 x_1 + k_2 x_2 + k_3 x_3, \qquad \mathbf{C} = \begin{pmatrix} C_1 \\ C_2 \\ C_3 \\ C_4 \end{pmatrix}. \qquad (5)$$

Inserting the ansatz (plane wave)

$$\Psi(\mathbf{x}, t) = \mathbf{C} \exp(i(\mathbf{k} \cdot \mathbf{x} - \omega t)) \qquad (6)$$

into the Dirac equation (1) we obtain the linear equation

$$\begin{pmatrix} -im_0 c - \hbar\omega/c & 0 & -i\hbar k_3 & -i\hbar k_1 - \hbar k_2 \\ 0 & -icm_0 - \hbar\omega/c & -i\hbar k_1 + \hbar k_2 & i\hbar k_3 \\ i\hbar k_3 & i\hbar k_1 + \hbar k_2 & \hbar\omega/c - icm_0 & 0 \\ i\hbar k_1 - \hbar k_2 & -i\hbar k_3 & 0 & \hbar\omega/c - icm_0 \end{pmatrix} \begin{pmatrix} C_1 \\ C_2 \\ C_3 \\ C_4 \end{pmatrix} = \begin{pmatrix} 0 \\ 0 \\ 0 \\ 0 \end{pmatrix},$$

$$(7)$$

where we used the fact that for $j = 1, 2, 3$

$$\frac{\partial}{\partial x_j} \Psi = \mathbf{C} i k_j \exp(i(\mathbf{k} \cdot \mathbf{x} - \omega t)) \qquad (8a)$$

and

$$\frac{\partial}{\partial x_4} \Psi = \mathbf{C}(-i\omega/c) \exp(i(\mathbf{k} \cdot \mathbf{x} - \omega t)). \qquad (8b)$$

From (7) we obtain the dispersion law (also called dispersion relation)

$$E^2 = c^2 \mathbf{p}^2 + m^2 c^4, \qquad (9)$$

where

$$E = \hbar\omega, \qquad p_j = \hbar k_j. \qquad (10)$$

In the program we insert ansatz (6) into (1) to find the linear equation (7). Finally we determine the dispersion law.

```
%dirac.red;

matrix g1(4,4), g2(4,4), g3(4,4), g4(4,4);
g1 := -i*hb*mat((0,0,0,-i),(0,0,-i,0),(0,i,0,0),(i,0,0,0));
g2 := -i*hb*mat((0,0,0,-1),(0,0,1,0),(0,1,0,0),(-1,0,0,0));
g3 := -i*hb*mat((0,0,-i,0),(0,0,0,i),(i,0,0,0),(0,-i,0,0));
g4 := -i*hb*mat((1,0,0,0),(0,1,0,0),(0,0,-1,0),(0,0,0,-1));
matrix psi(4,1);
psi := mat((u1),(u2),(u3),(u4));
matrix p1(4,1), p2(4,1), p3(4,1), p4(4,1);
p1 := g1*psi; p2 := g2*psi; p3 := g3*psi; p4 := g4*psi;
depend u1, x1, x2, x3, x4; depend u2, x1, x2, x3, x4;
depend u3, x1, x2, x3, x4; depend u4, x1, x2, x3, x4;
p1 := sub({u1=df(u1,x1),u2=df(u2,x1),u3=df(u3,x1),u4=df(u4,x1)},p1);
p2 := sub({u1=df(u1,x2),u2=df(u2,x2),u3=df(u3,x2),u4=df(u4,x2)},p2);
p3 := sub({u1=df(u1,x3),u2=df(u2,x3),u3=df(u3,x3),u4=df(u4,x3)},p3);
p4 := sub({u1=df(u1,x4),u2=df(u2,x4),u3=df(u3,x4),u4=df(u4,x4)},p4);

u1 := C1*exp(i*(k1*x1+k2*x2+k3*x3-om*x4/c));
u2 := C2*exp(i*(k1*x1+k2*x2+k3*x3-om*x4/c));
u3 := C3*exp(i*(k1*x1+k2*x2+k3*x3-om*x4/c));
u4 := C4*exp(i*(k1*x1+k2*x2+k3*x3-om*x4/c));

psi := p1 + p2 + p3 + p4 - i*m*c*psi;

matrix w(4,1);
w := psi/(exp(i*(k1*x1+k2*x2+k3*x3-om*x4/c)));
matrix A(4,4);
A(1,1) := coeffn(w(1,1),C1,1); A(1,2) := coeffn(w(1,1),C2,1);
A(1,3) := coeffn(w(1,1),C3,1); A(1,4) := coeffn(w(1,1),C4,1);
A(2,1) := coeffn(w(2,1),C1,1); A(2,2) := coeffn(w(2,1),C2,1);
A(2,3) := coeffn(w(2,1),C3,1); A(2,4) := coeffn(w(2,1),C4,1);
A(3,1) := coeffn(w(3,1),C1,1); A(3,2) := coeffn(w(3,1),C2,1);
A(3,3) := coeffn(w(3,1),C3,1); A(3,4) := coeffn(w(3,1),C4,1);
A(4,1) := coeffn(w(4,1),C1,1); A(4,2) := coeffn(w(4,1),C2,1);
A(4,3) := coeffn(w(4,1),C3,1); A(4,4) := coeffn(w(4,1),C4,1);
res := det(A);
```

The output is

```
P1 := MAT(( - HB*U4),( - HB*U3),(HB*U2),(HB*U1))$
P2 := MAT((I*HB*U4),( - I*HB*U3),( - I*HB*U2),(I*HB*U1))$
P3 := MAT(( - HB*U3),(HB*U4),(HB*U1),( - HB*U2))$
P4 := MAT(( - I*HB*U1),( - I*HB*U2),(I*HB*U3),(I*HB*U4))$
```

```
P1 := MAT(( - DF(U4,X1)*HB),( - DF(U3,X1)*HB),(DF(U2,X1)*HB),
(DF(U1,X1)*HB))$
P2 := MAT((DF(U4,X2)*I*HB),( - DF(U3,X2)*I*HB),( - DF(U2,X2)*I*HB),
(DF(U1,X2)*I*HB))$
P3 := MAT(( - DF(U3,X3)*HB),(DF(U4,X3)*HB),(DF(U1,X3)*HB),
( - DF(U2,X3)*HB))$
P4 := MAT(( - DF(U1,X4)*I*HB),( - DF(U2,X4)*I*HB),(DF(U3,X4)*I*HB),
(DF(U4,X4)*I*HB))$

W := MAT((( - (C**2*I*M*C1 + C*I*HB*K1*C4 + C*I*HB*K3*C3 +
C*HB*K2*C4 + HB*C1*OM))/C),(( - C**2*I*M*C2 - C*I*HB*K1*C3 +
C*I*HB*K3*C4 + C*HB*K2*C3 - HB*OM*C2)/C),(( - C**2*I*M*C3 +
C*I*HB*C1*K3 + C*I*HB*K1*C2 + C*HB*K2*C2 + HB*OM*C3)/C),
(( - C**2*I*M*C4 + C*I*HB*C1*K1 - C*I*HB*K3*C2 -
C*HB*C1*K2 + HB*OM*C4)/C))$


A(1,1) := ( - (C**2*I*M + HB*OM))/C$
A(1,2) := 0$
A(1,3) :=  - I*HB*K3$
A(1,4) :=  - HB*(I*K1 + K2)$
A(2,1) := 0$
A(2,2) := ( - (C**2*I*M + HB*OM))/C$
A(2,3) := HB*( - I*K1 + K2)$
A(2,4) := I*HB*K3$
A(3,1) := I*HB*K3$
A(3,2) := HB*(I*K1 + K2)$
A(3,3) := ( - C**2*I*M + HB*OM)/C$
A(3,4) := 0$
A(4,1) := HB*(I*K1 - K2)$
A(4,2) :=  - I*HB*K3$
A(4,3) := 0$
A(4,4) := ( - C**2*I*M + HB*OM)/C$


RES := (C**8*M**4 + 2*C**6*M**2*HB**2*K1**2 +
2*C**6*M**2*HB**2*K2**2 + 2*C**6*M**2*HB**2*K3**2 +
2*C**4*M**2*HB**2*OM**2 + C**4*HB**4*K1**4 +
2*C**4*HB**4*K1**2*K2**2 + 2*C**4*HB**4*K1**2*K3**2 +
C**4*HB**4*K2**4 + 2*C**4*HB**4*K2**2*K3**2 +
C**4*HB**4*K3**4 + 2*C**2*HB**4*K1**2*OM**2 +
2*C**2*HB**4*K2**2*OM**2 + 2*C**2*HB**4*K3**2*OM**2 +
HB**4*OM**4)/C**4$
```

## 35. Perturbation Theory

Let

$$\hat{H}_\epsilon = \hat{H}_0 + \epsilon \hat{V} \tag{1}$$

be a Hamilton operator with discrete spectrum for $\epsilon \geq 0$. Assume further that the eigenvalues are not degenerate. Eigenfunctions are assumed to be real orthonormal. Furthermore the eigenvalues and eigenfunctions of $\hat{H}_0$ are known. Finally it is assumed that the expansion does not diverge.

*Remark:* Degeneracies of eigenvalues are in general related to symmetries of the Hamilton operator $\hat{H}$. If the Hamilton operator $\hat{H}$ admits discrete symmetries the Hilbert space can be decomposed into invariant subspaces with respect to $\hat{H}$. These invariant subspaces are again Hilbert spaces and the perturbation expansion can be performed in these subspaces.

Our goal is to calculate the eigenvalues $E_n$ as a function of $\epsilon$. We define

$$E_n(\epsilon) : \text{ eigenvalues of } \hat{H}_\epsilon, \qquad |u_n(\epsilon)\rangle : \text{ eigenfunctions of } \hat{H}_\epsilon$$

where we assume that the eigenfunctions form an orthonormal basis in the underlying Hilbert space. In the following we write $E_n$ and $|u_n\rangle$ instead of $E_n(\epsilon)$ and $|u_n(\epsilon)\rangle$, respectively. Furthermore we define

$$p_n := \langle u_n | \hat{V} | u_n \rangle, \qquad V_{mn} := \langle u_m | \hat{V} | u_n \rangle \quad (m \neq n). \tag{2}$$

In the following derivation we use the orthonormality and completeness relation, i.e.,

$$\langle u_m | u_n \rangle = \delta_{mn} \qquad \sum_{n \in I} |u_n\rangle\langle u_n| = I. \tag{3}$$

From

$$\hat{H}_\epsilon |u_n\rangle = (\hat{H}_0 + \epsilon \hat{V})|u_n\rangle = E_n |u_n\rangle \tag{4}$$

it follows that

$$\hat{H}_0 \frac{d|u_n\rangle}{d\epsilon} + \hat{V}|u_n\rangle + \epsilon \hat{V} \frac{d|u_n\rangle}{d\epsilon} = \frac{dE_n}{d\epsilon}|u_n\rangle + E_n \frac{d|u_n\rangle}{d\epsilon}. \tag{5}$$

Taking the scalar product with $\langle u_n|$ we obtain

$$\langle u_n|(\hat{H}_0 + \epsilon\hat{V})\frac{d|u_n\rangle}{d\epsilon} + \langle u_n|\hat{V}|u_n\rangle = \frac{dE_n}{d\epsilon} + E_n\langle u_n|\frac{d|u_n\rangle}{d\epsilon}, \tag{6}$$

where we have used $\langle u_n|u_n\rangle = 1$. Since $\langle u_n|E_n = \langle u_n|(\hat{H}_0 + \epsilon\hat{V})$ we find

$$\frac{dE_n}{d\epsilon} = p_n. \tag{7}$$

Taking the scalar product with $\langle u_m|$ we obtain for $(m \neq n)$,

$$\langle u_m|(\hat{H}_0 + \epsilon\hat{V})\frac{d|u_n\rangle}{d\epsilon} + \langle u_m|\hat{V}|u_n\rangle = E_n\langle u_m|\frac{d|u_n\rangle}{d\epsilon}, \tag{8}$$

where we have used $\langle u_m|u_n\rangle = 0$ $(m \neq n)$. Using (2) and

$$\langle u_m|(\hat{H}_0 + \epsilon\hat{V}) = \langle u_m|E_m \tag{9}$$

it follows that

$$V_{mn} = [E_n - E_m]\langle u_m|\frac{d|u_n\rangle}{d\epsilon}. \tag{10}$$

From the orthonormality relation (3) we find

$$\frac{d\langle u_m|}{d\epsilon}|u_n\rangle + \langle u_m|\frac{d|u_n\rangle}{d\epsilon} = 0. \tag{11}$$

Therefore

$$V_{mn} = [E_m - E_n]\frac{d\langle u_m|}{d\epsilon}|u_n\rangle. \tag{12}$$

Let us now calculate $dp_n/d\epsilon$. We obtain

$$\frac{dp_n}{d\epsilon} = \frac{d}{d\epsilon}\left(\langle u_n|\hat{V}|u_n\rangle\right) = \left(\frac{d}{d\epsilon}\langle u_n|\right)\hat{V}|u_n\rangle + \langle u_n|\hat{V}|\frac{d}{d\epsilon}|u_n\rangle. \tag{13}$$

Using the completeness relation it follows that

$$\frac{dp_n}{d\epsilon} = \sum_{m\in I}\left(\frac{d}{d\epsilon}\langle u_n|\right)|u_m\rangle V_{mn} + \sum_{m\in I}V_{nm}\langle u_m|\frac{d}{d\epsilon}|u_n\rangle. \tag{14}$$

We have

$$\langle u_n|\frac{d}{d\epsilon}|u_n\rangle + \left(\frac{d}{d\epsilon}\langle u_n|\right)|u_n\rangle = 0. \tag{15}$$

Inserting (10) yields

$$\frac{dp_n}{d\epsilon} = \sum_{m\neq n}\frac{V_{nm}V_{mn}}{E_n - E_m} + \sum_{m\neq n}\frac{V_{mn}V_{nm}}{E_n - E_m}. \tag{16}$$

Consequently,

$$\frac{dp_n}{d\epsilon} = 2\sum_{m\neq n}\frac{V_{mn}V_{nm}}{E_n - E_m}, \tag{17}$$

where $V_{mn} = V_{nm}$. Let us now calculate $dV_{mn}/d\epsilon$ where $m\neq n$. We obtain

$$\frac{dV_{mn}}{d\epsilon} = \frac{d}{d\epsilon}\langle u_m|\hat{V}|u_n\rangle = \left(\frac{d}{d\epsilon}\langle u_m|\right)\hat{V}|u_n\rangle + \langle u_m|\hat{V}|\frac{d}{d\epsilon}|u_n\rangle. \tag{18}$$

Using the completeness relation gives

$$\frac{dV_{mn}}{d\epsilon} = \sum_{\ell \in I} \left( \frac{d}{d\epsilon} \langle u_m | \right) |u_\ell\rangle \langle u_\ell | \hat{V} | u_n \rangle + \sum_{\ell \in I} \langle u_m | \hat{V} | u_\ell \rangle \langle u_\ell | \frac{d}{d\epsilon} | u_n \rangle \qquad (19)$$

or

$$\frac{dV_{mn}}{d\epsilon} = \sum_{\ell \neq m} \frac{V_{m\ell}}{E_m - E_\ell} \langle u_\ell | \hat{V} | u_n \rangle + \sum_{\ell \neq n} \langle u_m | \hat{V} | u_\ell \rangle \frac{V_{\ell n}}{E_n - E_\ell}. \qquad (20)$$

Consequently,

$$\frac{dV_{mn}}{d\epsilon} = \sum_{\ell \neq (m,n)} \left[ V_{m\ell} V_{\ell n} \left( \frac{1}{E_m - E_\ell} + \frac{1}{E_n - E_\ell} \right) \right] + \frac{V_{mn}}{E_m - E_n} (p_n - p_m). \qquad (21)$$

To summarize: We find the following autonomous system of first order ordinary differential equations

$$\frac{dE_n}{d\epsilon} = p_n, \qquad \frac{dp_n}{d\epsilon} = 2 \sum_{m \neq n} \frac{V_{mn}^2}{E_n - E_m}$$

$$\frac{dV_{mn}}{d\epsilon} = \sum_{\ell \neq (m,n)} \left[ V_{m\ell} V_{\ell n} \left( \frac{1}{E_m - E_\ell} + \frac{1}{E_n - E_\ell} \right) \right] + \frac{V_{mn}(p_n - p_m)}{E_m - E_n}. \qquad (22)$$

This system of differential equations must be solved together with the "initial conditions"

$$E_n(\epsilon = 0)$$

$$p_n(\epsilon = 0) := \langle u_n(\epsilon = 0) | \hat{V} | u_n(\epsilon = 0) \rangle$$

$$V_{mn}(\epsilon = 0) := \langle u_m(\epsilon = 0) | \hat{V} | u_n(\epsilon = 0) \rangle.$$

This means that the eigenvalues $(E_n(\epsilon = 0))$ and eigenfunctions of $\hat{H}_0$ $(u_n(\epsilon = 0))$ must be known.

In the program we consider the three level system $\hat{H}_\epsilon = \hat{H}_0 + \epsilon\hat{V}$, where

$$\hat{H}_0 = \begin{pmatrix} 0 & 0 & 0 \\ 0 & 1 & 0 \\ 0 & 0 & 2 \end{pmatrix}, \qquad \hat{V} = \begin{pmatrix} 0 & 1 & 0 \\ 1 & 0 & 1 \\ 0 & 1 & 0 \end{pmatrix}. \tag{23}$$

The initial conditions are $E_0(0) = 0$, $E_1(0) = 1$, $E_2(0) = 2$, $p_0(0) = 0$, $p_1(0) = 0$, $p_2(0) = 0$, $V_{01} = 1$, $V_{02} = 0$, $V_{12} = 1$. Notice that $V_{ij} = V_{ji}$.

```
%pertu.red;

operator V, Q, x, xs;
depend V(j), x(k);
depend Q(j), x(k);

V(1):=x(4);
V(2):=x(5);
V(3):=x(6);
V(4):=2*(x(7)*x(7)/(x(1)-x(2)) + x(8)*x(8)/(x(1)-x(3)));
V(5):=2*(x(7)*x(7)/(x(2)-x(1)) + x(9)*x(9)/(x(2)-x(3)));
V(6):=2*(x(8)*x(8)/(x(3)-x(1)) + x(9)*x(9)/(x(3)-x(2)));
V(7):=x(9)*x(8)*(1/(x(2)-x(3))+1/(x(1)-x(3)))+
x(7)*(x(4)-x(5))/(x(2)-x(1));
V(8):=x(7)*x(9)*(1/(x(3)-x(2))+1/(x(1)-x(2)))+
x(8)*(x(4)-x(6))/(x(3)-x(1));
V(9):=x(7)*x(8)*(1/(x(3)-x(1))+1/(x(2)-x(1)))+
x(9)*(x(5)-x(6))/(x(3)-x(2));

for j:=1:9 do
Q(j) := for k := 1:9 sum (V(k)*df(V(j),x(k)));.

on rounded$

count := 0.0; ep := 0.02; % ep is the step length;
xs(1) := 0.0; xs(2) := 1.0; xs(3) := 2.0; % initial conditions;
xs(4) := 0.0; xs(5) := 0.0; xs(6) := 0.0; % initial conditions;
xs(7) := 1.0; xs(8) := 0.0; xs(9) := 1.0; % initial conditions;
```

```
while count < 1.0 do
<< x(1) := xs(1); x(2) := xs(2); x(3) := xs(3);
x(4) := xs(4); x(5) := xs(5); x(6) := xs(6);
x(7) := xs(7); x(8) := xs(8); x(9) := xs(9);
xs(1) := x(1) + ep*(V(1) + ep/2*Q(1));
xs(2) := x(2) + ep*(V(2) + ep/2*Q(2));
xs(3) := x(3) + ep*(V(3) + ep/2*Q(3));
xs(4) := x(4) + ep*(V(4) + ep/2*Q(4));
xs(5) := x(5) + ep*(V(5) + ep/2*Q(5));
xs(6) := x(6) + ep*(V(6) + ep/2*Q(6));
xs(7) := x(7) + ep*(V(7) + ep/2*Q(7));
xs(8) := x(8) + ep*(V(8) + ep/2*Q(8));
xs(9) := x(9) + ep*(V(9) + ep/2*Q(9));
count := count + ep>>$

write "count = ", count;
E0 := xs(1); E1 := xs(2); E2 := xs(3);
p0 := xs(4); p1 := xs(5); p2 := xs(6);
U01 := xs(7); U02 := xs(8); U12 := xs(9);
```

The output is

```
count = 1$
e0 :=  - 0.732138253349$
e1 := 1$
e2 := 2.73213825335$
p0 :=  - 1.15447985106$
p1 := 0$
p2 := 1.15447985106$
u01 := 0.577268440498$
u02 := 1.04663624657E-17$
u12 := 0.577268440498$
```

*Remark:* The exact solutions are

$$E_0(\epsilon = 1) = -(\sqrt{3} - 1), \qquad E_1(\epsilon = 1) = 1, \qquad E_2(\epsilon = 1) = \sqrt{3} + 1.$$

## 36. Elastic Scattering

Consider the problem of elastic scattering. The asymptotic boundary condition

$$u(r, \theta, \phi) \rightarrow e^{ikz} + f(\theta)\frac{e^{ikr}}{r} \quad \text{for} \quad kr \rightarrow \infty \tag{1}$$

is imposed on the wave function of a scattering problem (the so-called *Sommerfeld radiation condition*), where $z = r\cos\theta$. This leads to a superposition of incident and scattered currents without noticeable interference at large distances from the scattering object. We derive the relation between scattering amplitudes and cross section. The current density is defined by

$$\mathbf{s} := \frac{\hbar}{2im}N(u^*\nabla u - u\nabla u^*) \tag{2}$$

with a normalization constant $N$ determining the absolute intensity of the current. The gradient $\nabla$, in spherical polar coordinates, has the components

$$\nabla_r := \frac{\partial}{\partial r}, \qquad \nabla_\theta := \frac{1}{r}\frac{\partial}{\partial\theta}, \qquad \nabla_\phi := \frac{1}{r\sin\theta}\frac{\partial}{\partial\phi}. \tag{3}$$

Since $u$ does not depend upon $\phi$ we have $s_\phi = 0$. For the two other components of $\mathbf{s}$ we obtain

$$\frac{s_r}{N} = \frac{\hbar k}{m}\left(\cos\theta + \frac{|f|^2}{r^2}\right)$$

$$+\frac{\hbar}{2m}\left((kr(1+\cos\theta)+i)\frac{e^{ik(r-z)}}{r^2}f + (kr(1+\cos\theta)-i)\frac{e^{ik(z-r)}}{r^2}f^*\right) \tag{4a}$$

$$\frac{s_\theta}{N} = -\frac{\hbar k}{m}\sin\theta + \frac{\hbar}{2mi}\left((f' - ikr\sin\theta f)\frac{e^{ik(r-z)}}{r^2} - (f^{*\prime} + ikr\sin\theta f^*)\frac{e^{ik(z-r)}}{r^2}\right)$$

$$+\frac{\hbar}{2mir^3}(f'f^* - ff^{*\prime}). \tag{4b}$$

Here the prime denotes differentiation with respect to $\theta$. The equations hold only asymptotically for $kr \to \infty$. Thus we can neglect the term proportional to $r^{-3}$ for $s_\theta$. Moreover, in the big bracket we can neglect the terms proportional to $r^{-2}$. Thus we obtain the simpler expressions,

$$\frac{s_r}{N} = \frac{\hbar k}{m}\left(\cos\theta + \frac{|f|^2}{r^2} + \frac{1+\cos\theta}{2r}(fe^{ik(r-z)} + f^*e^{ik(z-r)})\right) \tag{5a}$$

$$\frac{s_\theta}{N} = \frac{\hbar k}{m}\left\{-\sin\theta - \frac{\sin\theta}{2r}(fe^{ik(r-z)} + f^*e^{ik(z-r)})\right\}. \tag{5b}$$

In these expressions the first terms are independent of $r$. They represent the decomposition into $r$ and $\theta$ components of the plane wave contribution with the current density $s_0 = \hbar k N/m$ in $z$ direction. Since $\hbar k/m = v$ is the particle velocity, $N$ turns out to be the number of particles per unit volume in the incident current. The second term of $s_r$ is the radial current which we can identify as the scattered intensity. In order to obtain a finite intensity, we have to use some detector of a finite small solid angle $\delta\Omega$ at a distance $r$ through whose surface $(r^2\delta\Omega)$ a scattered current of

$$\frac{\delta S}{N} = \int_{\delta\Omega}\left(\frac{\hbar k}{m}\frac{|f|^2}{r^2}\right)r^2d\Omega = \frac{\hbar k}{m}\int_{\delta\Omega}|f|^2d\Omega \tag{6}$$

particles per second will pass and be counted, where

$$d\Omega := \sin^2\theta d\theta d\phi. \tag{7}$$

If this current is divided by the plane wave current density of $(\hbar k/m)N$ particles per second and cm$^2$, a quantity

$$\delta\sigma = \int_{\delta\Omega}|f(\theta)|^2d\Omega \simeq |f(\theta)|^2\delta\Omega \tag{8}$$

independent of the primary intensity, and thus characteristic of the scattering properties of the scattering interaction, will emerge. This ratio has the dimension of area and is called the *differential cross section*.

In the program we insert (1) into (2) to obtain $sr$ and $st$ (4a and 4b). Then we derive (5b). The notation is: $u \to u$, $u^* \to uc$, $f \to f$, and $f^* \to fc$.

```
%escatt.red;

depend u, r, theta;
depend uc, r, theta;
depend f, theta;
depend fc, theta;

u  := exp(i*k*r*cos(theta)) + f*exp(i*k*r)/r;
uc := sub({i=-i,f=fc},u);

sr := N*hb/(2*i*m)*(uc*df(u,r) - u*df(uc,r));

st := N*hb/(2*i*m)*(uc*(1/r)*df(u,theta) - u*(1/r)*df(uc,theta));

nst := num(st);  % numerator of st;
dnt := den(st);  % denominator of st;

sta := coeffn(nst,r,3)*r**3 + coeffn(nst,r,2)*r**2;

sta := sta/dnt;
```

The output is

```
UC := (E**(COS(THETA)*I*K*R)*FC + E**(I*K*R)*R)/
(E**(COS(THETA)*I*K*R + I*K*R)*R)$


SR := (N*HB*(E**(2*COS(THETA)*I*K*R)*COS(THETA)*I*K*R*FC +
E**(2*COS(THETA)*I*K*R)*I*K*R*FC + E**(2*COS(THETA)*I*K*R)*FC +
2*E**(COS(THETA)*I*K*R + I*K*R)*COS(THETA)*I*K*R**2 +
2*E**(COS(THETA)*I*K*R + I*K*R)*F*I*K*FC +
E**(2*I*K*R)*COS(THETA)*F*I*K*R + E**(2*I*K*R)*F*I*K*R -
E**(2*I*K*R)*F))/(2*E**(COS(THETA)*I*K*R +
I*K*R)*I*M*R**2)$


ST := (N*HB*( - E**(2*COS(THETA)*I*K*R)*DF(FC,THETA)*R -
E**(2*COS(THETA)*I*K*R)*SIN(THETA)*I*K*R**2*FC +
E**(COS(THETA)*I*K*R + I*K*R)*DF(F,THETA)*FC -
E**(COS(THETA)*I*K*R + I*K*R)*DF(FC,THETA)*F -
2*E**(COS(THETA)*I*K*R + I*K*R)*SIN(THETA)*I*K*R**3 +
E**(2*I*K*R)*DF(F,THETA)*R -
E**(2*I*K*R)*SIN(THETA)*F*I*K*R**2))/
(2*E**(COS(THETA)*I*K*R + I*K*R)*I*M*R**3)$


NST := N*HB*( - E**(2*COS(THETA)*I*K*R)*DF(FC,THETA)*R -
E**(2*COS(THETA)*I*K*R)*SIN(THETA)*I*K*R**2*FC +
E**(COS(THETA)*I*K*R + I*K*R)*DF(F,THETA)*FC -
E**(COS(THETA)*I*K*R + I*K*R)*DF(FC,THETA)*F -
2*E**(COS(THETA)*I*K*R + I*K*R)*SIN(THETA)*I*K*R**3 +
E**(2*I*K*R)*DF(F,THETA)*R - E**(2*I*K*R)*SIN(THETA)*F*I*K*R**2)$


DNT := 2*E**(COS(THETA)*I*K*R + I*K*R)*I*M*R**3$


STA := SIN(THETA)*I*K*N*R**2*HB*( - E**(2*COS(THETA)*I*K*R)*FC -
2*E**(COS(THETA)*I*K*R + I*K*R)*R - E**(2*I*K*R)*F)$


STA := (SIN(THETA)*K*N*HB*( - E**(2*COS(THETA)*I*K*R)*FC -
2*E**(COS(THETA)*I*K*R + I*K*R)*R - E**(2*I*K*R)*F))/
(2*E**(COS(THETA)*I*K*R + I*K*R)*M*R)$
```

## 37. Exceptional Points

Let $\hat{H}_0$ and $\hat{H}_1$ be a pair of real symmetric $N \times N$ matrices, where $\hat{H}_0$ is a diagonal matrix. Let

$$\hat{H}(\lambda) := \hat{H}_0 + \lambda \hat{H}_1. \tag{1}$$

When $\lambda$ is real, $\hat{H}(\lambda)$ is diagonalizable with eigenvalues $E_1(\lambda)$, $E_2(\lambda)$, ..., $E_N(\lambda)$. The eigenvalues are given by the characteristic polynomial

$$P(E, \lambda) := \det(\hat{H}(\lambda) - EI) = 0, \tag{2}$$

where $I$ is the $N \times N$ unit matrix. When $\lambda$ is complex, the eigenvalues may be viewed as the $N$ values of a single function $E(\lambda)$ of $\lambda$, analytic on a Riemann surface with $N$ sheets joined at branch point singularities in the complex plane. The *exceptional points* in the complex $\lambda$ plane are defined by the solution of (1) together with

$$\frac{d}{dE} \det(\hat{H}(\lambda) - EI) = 0. \tag{3}$$

In the program we consider a two level system, namely

$$\hat{H}(\lambda) = \begin{pmatrix} 0 & 0 \\ 0 & 1 \end{pmatrix} + \lambda \begin{pmatrix} 0 & 1 \\ 1 & 0 \end{pmatrix}. \tag{4}$$

From (2) and (3) we find that

$$E^2 - E - \lambda^2 = 0, \qquad 2E - 1 = 0. \tag{5}$$

Thus the exceptional points are

$$\lambda_1 = \frac{i}{2}, \qquad \lambda_2 = -\frac{i}{2}. \tag{6}$$

In the program we first calculate (5) from the $2 \times 2$ matrices $\hat{H}_0$ and $\hat{H}_1$. Then we determine the exceptional points by solving (5) with respect to $\lambda$.

```
%except.red;

matrix H0, H1, ID;
H0 := mat((0,0),(0,1));
H1 := mat((0,lam),(lam,0));
ID := mat((1,0),(0,1));
res1 := det(H0+H1-En*ID);
res2 := df(res1,En);

list := solve(res2=0,En);
part(list,1);
En := part(ws,2);
res3 := res1;
res4 := solve(res3=0,lam);
```

The output is

```
RES1 :=  - LAM**2 + EN**2 - EN$

RES2 := 2*EN - 1$

LIST := {EN=1/2}$

EN := 1/2$

RES3 := ( - 4*LAM**2 - 1)/4$

RES4 := {LAM= - 1/2*I,LAM=1/2*I}$
```

## 38. Expansion of $\exp(L)A\exp(-L)$

Let $A$ and $L$ be two $n \times n$ matrices. We evaluate

$$e^L A e^{-L}. \tag{1}$$

To evaluate (1) we consider

$$A(\epsilon) := e^{\epsilon L} A e^{-\epsilon L}, \tag{2}$$

where $\epsilon$ is a real parameter. Taking the derivative of $A(\epsilon)$ with respect to $\epsilon$ yields

$$\frac{dA}{d\epsilon} = L e^{\epsilon L} A e^{-\epsilon L} - e^{\epsilon L} A e^{-\epsilon L} L = [L, A(\epsilon)]. \tag{3}$$

The second derivative gives

$$\frac{d^2 A}{d\epsilon^2} = \left[ L, \frac{dA}{d\epsilon} \right] = [L, [L, A(\epsilon)]] \tag{4}$$

and so on. Consequently, we can write the matrix $e^L A e^{-L} = A(1)$ as a Taylor series expansion about the origin, i.e.,

$$A(1) = A(0) + \frac{1}{1!}\frac{dA(0)}{d\epsilon} + \frac{1}{2!}\frac{d^2 A(0)}{d\epsilon^2} + \cdots, \tag{5}$$

where $A(\epsilon = 0) = A$ and

$$\frac{dA(0)}{d\epsilon} := \left. \frac{dA(\epsilon)}{d\epsilon} \right|_{\epsilon=0}. \tag{6}$$

Thus we find that

$$e^L A e^{-L} = A + [L, A] + \frac{1}{2!}[L, [L, A]] + \frac{1}{3!}[L, [L, [L, A]]] + \cdots . \tag{7}$$

In the program we calculate the right-hand side of (7).

```
%expan.red;

operator A, B, L, expon;
noncom A, B, L, expon;

B(1) := expon(ep*L(0))*A(0)*expon(-ep*L(0));

let df(expon(ep*L(0)),ep) = L(0)*expon(ep*L(0));
let df(expon(-ep*L(0)),ep) = -expon(-ep*L(0))*L(0);

result := for n := 0:4 sum (1/factorial(n))*df(B(1),ep,n);

ep := 0;

let expon(0) = 1;

result;
```

The output is

```
RESULT := (L(0)**4*EXPON(L(0)*EP)*A(0)*EXPON( - L(0)*EP) -
4*L(0)**3*EXPON(L(0)*EP)*A(0)*EXPON( - L(0)*EP)*L(0) +
4*L(0)**3*EXPON(L(0)*EP)*A(0)*EXPON( - L(0)*EP) +
6*L(0)**2*EXPON(L(0)*EP)*A(0)*EXPON( - L(0)*EP)*L(0)**2 -
12*L(0)**2*EXPON(L(0)*EP)*A(0)*EXPON( - L(0)*EP)*L(0) +
12*L(0)**2*EXPON(L(0)*EP)*A(0)*EXPON( - L(0)*EP) -
4*L(0)*EXPON(L(0)*EP)*A(0)*EXPON( - L(0)*EP)*L(0)**3 +
12*L(0)*EXPON(L(0)*EP)*A(0)*EXPON( - L(0)*EP)*L(0)**2 -
24*L(0)*EXPON(L(0)*EP)*A(0)*EXPON( - L(0)*EP)*L(0) +
24*L(0)*EXPON(L(0)*EP)*A(0)*EXPON( - L(0)*EP) +
EXPON(L(0)*EP)*A(0)*EXPON( - L(0)*EP)*L(0)**4 -
4*EXPON(L(0)*EP)*A(0)*EXPON( - L(0)*EP)*L(0)**3 +
12*EXPON(L(0)*EP)*A(0)*EXPON( - L(0)*EP)*L(0)**2 -
24*EXPON(L(0)*EP)*A(0)*EXPON( - L(0)*EP)*L(0) +
24*EXPON(L(0)*EP)*A(0)*EXPON( - L(0)*EP))/24$

EP := 0$
let expon(0) = 1;

(A(0)*L(0)**4 - 4*A(0)*L(0)**3 + 12*A(0)*L(0)**2 - 24*A(0)*L(0) +
24*A(0) + L(0)**4*A(0) - 4*L(0)**3*A(0)*L(0) + 4*L(0)**3*A(0) +
6*L(0)**2*A(0)*L(0)**2 - 12*L(0)**2*A(0)*L(0) + 12*L(0)**2*A(0) -
4*L(0)*A(0)*L(0)**3 + 12*L(0)*A(0)*L(0)**2 - 24*L(0)*A(0)*L(0) +
24*L(0)*A(0))/24$
```

In the following two REDUCE programs we consider the matrices

$$A = \begin{pmatrix} 2 & 3 \\ 3 & 4 \end{pmatrix}, \qquad L = \begin{pmatrix} 2 & 3 \\ 4 & 5 \end{pmatrix}.$$

In the program we calculate the right hand side of (7) up to order $n = 4$ for the matrices $A$ and $L$.

```
%expan1.red;

matrix R(2,2), A(2,2), L(2,2), Temp(2,2);

A := Mat((2,3),(3,4));
L := Mat((2,3),(4,5));

n := 4;

R := A;
Temp := A;
for k := 1:n do <<
    Temp := (L*Temp - Temp*L)/k;
    R := R + Temp;
>>;

R := R;
```

The output is

```
R := MAT((109/8,393/8),(( - 375)/8,( - 61)/8))$
```

```
%expan2.red;

matrix R(2,2), A(2,2), L(2,2);
matrix Temp(2,2), Temp1(2,2), Temp2(2,2), Id(2,2);

A  := Mat((2,3),(3,4));
L  := Mat((2,3),(4,5));
Id := Mat((1,0),(0,1));

define n = 4;

Temp1 := Id;
Temp2 := Id;
Temp  := Id;

for k := 1:n do <<
    Temp  := Temp*L*ep/k;
    Temp1 := Temp1 + Temp;
    Temp2 := Temp2 + (-1)**k*Temp;
    >>;

let ep**(n+1) = 0;

R  := Temp1*A*Temp2;
ep := 1;
R  := R;
```

The output is

```
R := MAT(((3*(285*EP**4 - 228*EP**3 + 60*EP**2 - 24*EP + 16))/24,
(3*(171*EP**4 - 76*EP**3 + 36*EP**2 - 8*EP + 8))/8),
(( - 399*EP**4 + 76*EP**3 - 84*EP**2 + 8*EP + 24)/8,
(3*( - 285*EP**4 + 228*EP**3 - 60*EP**2 + 24*EP + 32))/24))$

R := MAT((109/8,393/8),(( - 375)/8,( - 61)/8))$
```

## 39. Expansion of $(A - \epsilon B)^{-1}$

Let $A$ and $B$ be $n \times n$ matrices. Assume that $A^{-1}$ and $(A - \epsilon B)^{-1}$ exist. We evaluate the expansion of

$$(A - \epsilon B)^{-1} \tag{1}$$

as a power series in $\epsilon$, where $\epsilon$ is a real parameter. We set

$$(A - \epsilon B)^{-1} := \sum_{n=0}^{\infty} \epsilon^n L_n. \tag{2}$$

Multiplying (2) on the left by $(A - \epsilon B)$ we obtain

$$I = \sum_{n=0}^{\infty} \epsilon^n (A - \epsilon B) L_n = A L_0 + \sum_{n=1}^{\infty} \epsilon^n (A L_n - B L_{n-1}), \tag{3}$$

where $I$ is the $n \times n$ unit matrix. By equating coefficients of powers of $\epsilon$ we find that

$$L_0 = A^{-1} \tag{4}$$

and

$$L_n = A^{-1} B L_{n-1}, \tag{5}$$

where $n = 1, 2, \ldots$ . Thus (5) defines a recursion relation. Consequently,

$$(A - \epsilon B)^{-1} = A^{-1} + \epsilon A^{-1} B A^{-1} + \epsilon^2 A^{-1} B A^{-1} B A^{-1} + \cdots \tag{6}$$

We give three implementations. In the second and third program we consider the matrices

$$A = \begin{pmatrix} 1 & 2 \\ 2 & 3 \end{pmatrix}, \qquad B = \begin{pmatrix} 2 & 3 \\ 3 & 4 \end{pmatrix}. \tag{7}$$

```
%expinv.red;

operator L, B, AI;
noncom L, B, AI;

L(0) := AI(0);
for j := 1:4 do
L(j) := AI(0)*B(0)*L(j-1);

for j := 0:4 do
write L(j);

result := for n := 0:4 sum ep^n*L(n);
```

The output is

```
AI(0)$
AI(0)*B(0)*AI(0)$
AI(0)*B(0)*AI(0)*B(0)*AI(0)$
AI(0)*B(0)*AI(0)*B(0)*AI(0)*B(0)*AI(0)$
AI(0)*B(0)*AI(0)*B(0)*AI(0)*B(0)*AI(0)*B(0)*AI(0)$


RESULT := AI(0)*B(0)*AI(0)*B(0)*AI(0)*B(0)*AI(0)*B(0)*AI(0)*EP**4 +
AI(0)*B(0)*AI(0)*B(0)*AI(0)*B(0)*AI(0)*EP**3 +
AI(0)*B(0)*AI(0)*B(0)*AI(0)*EP**2 + AI(0)*B(0)*AI(0)*EP + AI(0)$
```

```
%expinv1.red;

n := 5;

matrix A(2,2), B(2,2), R1(2,2), R2(2,2);

A := mat((1,2),(2,3));
B := mat((2,3),(3,4));

R1 := 1/A;
R2 := R1;

for j := 1 step 1 until n do
begin
R1 := R1*ep*B*A^(-1);
R2 := R2 + R1;
end;

R2 := R2;
```

The output is

```
R1 := MAT((-3,2),(2,-1))$

R2 := MAT((-3,2),(2,-1))$

R2 := MAT((2*ep**5 + ep**4 - ep**2 - 2*ep - 3,
- 3*ep**5 - 2*ep**4 - ep**3 + ep + 2),
( - 3*ep**5 - 2*ep**4 - ep**3 + ep + 2,
4*ep**5 + 3*ep**4 + 2*ep**3 + ep**2 - 1))$
```

```
%expinv2.red;
n := 5;

matrix A(2,2), B(2,2), R1(2,2), R2(2,2);

A := mat((1,2),(2,3));
B := mat((2,3),(3,4));

R1 := (A-ep*B)^(-1);

let ep = 0;
R2 := R1;
clear ep;

for i := 1:2 do
begin
for j := 1:2 do
begin
x := R1(i,j);
for k := 1:n do
begin
x := df(x,ep)/k;
let ep = 0;
y := x;
clear ep;
R2(i,j) := R2(i,j) + y*ep^k;
end;
end;
end;

R2 := R2;
```

The output is

```
R1 := MAT(((4*ep - 3)/(ep**2 - 2*ep + 1),( - 3*ep + 2)/
(ep**2 - 2*ep + 1)),(( - 3*ep + 2)/(ep**2 - 2*ep + 1),
(2*ep -1)/(ep**2 - 2*ep + 1)))$


R2 := MAT((-3,2),(2,-1))$

R2 := MAT((2*ep**5 + ep**4 - ep**2 - 2*ep - 3, - 3*ep**5 -
2*ep**4 - ep**3 + ep + 2),( - 3*ep**5 - 2*ep**4 - ep**3 +
ep + 2,4*ep**5 + 3*ep**4 + 2*ep**3 + ep**2 - 1))$
```

## 40. Heavyside-, Sign- and Delta Function

The *Heavyside function* is defined as

$$h(x) := \begin{cases} 1 & \text{for } x \geq 0 \\ 0 & \text{for } x < 0. \end{cases} \tag{1}$$

The *sign function* is defined as

$$\text{sgn(x)} := \begin{cases} 1 & \text{for } x > 0 \\ 0 & \text{for } x = 0 \\ -1 & \text{for } x < 0. \end{cases} \tag{2}$$

The *delta function* is defined as (in the sense of generalized functions)

$$(\delta(x), \phi(x)) := \phi(0), \tag{3}$$

where $\phi$ is a so-called test function. The delta function is a linear functional, i.e.

$$(\delta(x), \phi_1(x) + \phi_2(x)) = \phi_1(0) + \phi_2(0), \tag{4}$$

where $\phi_1$ and $\phi_2$ are test functions. The delta function is also the derivative (in the sense of generalized functions) of the Heavyside function.

In the program we implement the three functions and give some applications. The Heavyside function and and the sign function are implemented as procedures.

```
%heavy.red;

procedure h(x);
begin
  if x >= 0 then return 1 else return 0;
end;

procedure sign(x);
begin
  if x < 0 then return -1;
  if x = 0 then return 0;
  if x > 0 then return 1;
end;

operator d, f;
for all x let d*f(x) = f(0);
for all x let d*sin(x) = sin(0);
for all x let d*cos(x) = cos(0);

d*f(x);  d*sin(x);  d*cos(2*y);

h(1); h(0); h(-1);

sign(0); sign(2); sign(-10);

r1 := h(2)*sub(x=2,int(1,x)) - h(0)*sub(x=0,int(1,x));
r2 := sign(2)*sub(x=2,int(x^2,x)) - sign(-2)*sub(x=-2,int(x^2,x));
```

The output is

```
F(0)$
0$
1$

1$
1$
0$

0$
1$
-1$

R1 := 2$
R2 := 0$
```

## 41. Legendre Polynomials

The *Legendre polynomials* are defined by

$$P_0(x) := 1, \qquad P_n(x) := \frac{1}{2^n n!} \frac{d^n}{dx^n} (x^2 - 1)^n \tag{1}$$

where $n = 1, 2, \ldots$ (Rodrigues' formula). Since

$$P_n(x) = \frac{2^n}{n!} \frac{d^n}{dx^n} \left\{ \left( \frac{x-1}{2} \right)^n \left( \frac{x+1}{2} \right)^n \right\} \tag{2}$$

and applying the well-known formula for the $n$-th derivative of a product, we obtain

$$P_n(x) = \sum_{k=0}^{n} \binom{n}{k}^2 \left( \frac{x-1}{2} \right)^{n-k} \left( \frac{x+1}{2} \right)^k. \tag{3}$$

These polynomials obey the following linear second order differential equation

$$\frac{d}{dx} \left( (1 - x^2) \frac{dP_n}{dx} \right) + n(n+1) P_n = 0. \tag{4}$$

We find that

$$\int_{-1}^{+1} P_m(x) P_n(x) dx = \frac{2}{2n+1} \delta_{mn}. \tag{5}$$

Furthermore we have

$$(n+1) P_{n+1}(x) = (2n+1) x P_n(x) - n P_{n-1}(x) \tag{6}$$

with $n = 1, 2, \ldots$ and $P_0(x) = 1$ and $P_1(x) = x$. The first few Legendre polynomials are

$$P_0(x) = 1, \quad P_1(x) = x, \quad P_2(x) = \frac{1}{2}(3x^2 - 1), \quad P_3(x) = \frac{1}{2}(5x^2 - 3x). \tag{7}$$

In the program we implement recursion relation (6) as a procedure.

```
%Legend.red;

procedure leg(n,y);
if n=0 then 1
else
if n=1 then y
else
((2*n-1)*y*leg(n-1,y)+(1-n)*leg(n-2,y))/n;

r1 := leg(7,x);

%leg(7,x) satisfies (4);
r2 := df((1-x**2)*df(r1,x),x) + 7*(7+1)*r1;

%evaluation of (5) with n=m=7;
r3 := sub(x=1,int(r1*r1,x))-sub(x=-1,int(r1*r1,x));
```

The output is

```
r1 := (x*(429*x**6 - 693*x**4 + 315*x**2 - 35))/16$

r2 := 0$

r3 := 2/15$
```

## 42. Laguerre Polynomials

The *Laguerre polynomials* are defined by

$$L_n(x) := e^x \frac{d^n}{dx^n}(e^{-x}x^n), \qquad n = 0, 1, \ldots \tag{1}$$

where $L_0(x) := 1$. We find that $L_n$ satisfies the linear second order differential equation

$$x\frac{d^2 L_n}{dx^2} + (1 - x)\frac{dL_n}{dx} + nL_n = 0. \tag{2}$$

Furthermore we find the recursion formula

$$L_{n+1}(x) = (2n + 1 - x)L_n(x) - n^2 L_{n-1}(x), \tag{3}$$

where $L_0(x) = 1$ and $L_1(x) = 1 - x$. The Laguerre polynomials can also be defined by a generating function

$$(1 - t)^{-1} \exp(-xt/(1 - t)) = \sum_{n=0}^{\infty} \frac{L_n(x)t^n}{n!}. \tag{4}$$

The first few polynomials are

$$L_0(x) = 1, \quad L_1(x) = 1 - x, \quad L_2(x) = 2 - 4x + x^2. \tag{5}$$

The orthogonality relation of the Laguerre polynomials is given by

$$\int_0^{\infty} e^{-x} L_m(x)L_n(x)dx = 0, \qquad n \neq m \tag{6a}$$

$$\int_0^{\infty} e^{-x}(L_n(x))^2 dx = (n!)^2. \tag{6b}$$

In the program we implement recursion relation (3).

```
%laguer.red;

procedure lag(n,y);
if n=0 then 1
else
if n=1 then 1 - y
else
(2*n - 1 - y)*lag(n-1,y) - (n-1)**2*lag(n-2,y);

r1 := lag(7,x);

%r1 satisfies (2) with n = 7;
r2 := x*df(r1,x,2) + (1-x)*df(r1,x) + 7*r1;

%evaluation of (6b) for n = 7;
r3 := -sub(x=0,int(exp(-x)*r1*r1,x));
```

The output is

```
R1 :=  - X**7 + 49*X**6 - 882*X**5 + 7350*X**4 - 29400*X**3 +
52920*X**2 - 35280*X + 5040$

R2 := 0$

R3 := 25401600$
```

## 43. Hermite Polynomials

The *Hermite polynomials* are defined by

$$H_n(x) := (-1)^n \exp(x^2) \frac{d^n}{dx^n} \exp(-x^2), \tag{1}$$

where $n = 0, 1, 2, \ldots$ and $H_0(x) := 1$. They satisfy the linear differential equation

$$\frac{d^2 H_n}{dx^2} - 2x \frac{dH_n}{dx} + 2n H_n = 0. \tag{2}$$

We find that

$$\int\limits_{-\infty}^{\infty} \exp(-x^2) H_n(x) H_m(x) dx = 2^n n! \sqrt{\pi} \delta_{nm}, \tag{3}$$

where $n, m = 0, 1, 2, \ldots$ and $\delta_{nm}$ is the Kronecker delta. The recursion formula takes the form

$$H_{n+1}(x) = 2x H_n(x) - 2n H_{n-1}(x), \tag{4}$$

where $H_0(x) = 1$ and $H_1(x) = 2x$. The first few Hermite polynomials are given by

$$H_0(x) = 1, \quad H_1(x) = 2x, \quad H_2(x) = 4x^2 - 2, \quad H_3(x) = 8x^3 - 12x. \tag{5}$$

The Hermite polynomials can be defined from a generating function

$$\exp(2tx - t^2) = \sum_{n=0}^{\infty} \frac{H_n(x) t^n}{n!}. \tag{6}$$

Furthermore, we have the relation

$$\frac{dH_n(x)}{dx} = 2n H_{n-1}(x). \tag{7}$$

In the program we implement the recursion relation (4) as a procedure.

```
%hermite.red;

procedure her(n,y);
if n=0 then 1
else
if n=1 then 2*y
else
2*y*her(n-1,y) - 2*(n-1)*her(n-2,y);

r1 := her(7,x);

%r1 satisfies (2) with n = 7;
r2 := df(r1,x,2) - 2*x*df(r1,x) + 2*7*r1;

%we confirm (7) for n = 6;
r3 := df(her(6,x),x) - 2*6*her(5,x);

%implementation of (1) for n = 7;
n := 7;
r4 := (-1)**n*exp(x*x)*df(exp(-x*x),x,n);
```

The output is

```
R1 := 16*X*(8*X**6 - 84*X**4 + 210*X**2 - 105)$

R2 := 0$

R3 := 0$

R4 := 16*X*(8*X**6 - 84*X**4 + 210*X**2 - 105)$
```

## 44. Chebyshev Polynomials

The *Chebyshev polynomials* are defined by the relation

$$T_n(x) := \cos(n \arccos x), \qquad n = 0, 1, 2, \dots . \tag{1}$$

Thus $T_0 = 1$. Moreover, we have $T_{-n}(x) = T_n(x)$. From trigonometric formulas we find that

$$T_{n+m}(x) + T_{n-m}(x) = 2T_n(x)T_m(x). \tag{2}$$

For $m = 1$ we obtain the recursion relation

$$T_{n+1}(x) = 2xT_n(x) - T_{n-1}(x), \tag{3}$$

where $T_0(x) = 1$ and $T_1(x) = x$. Thus we can successively compute all $T_n(x)$. Let

$$x := \cos \theta. \tag{4}$$

Then

$$y = T_n(x) = \cos(n\theta). \tag{5}$$

Furthermore

$$\frac{dy}{dx} = \frac{n \sin(n\theta)}{\sin \theta} \tag{6}$$

and

$$\frac{d^2y}{dx^2} = \frac{-n^2 \cos(n\theta) + n \sin(n\theta) \cot \theta}{\sin^2 \theta} = -\frac{n^2 y}{1 - x^2} + \frac{x}{1 - x^2} \frac{dy}{dx}. \tag{7}$$

Thus the polynomials $T_n(x)$ satisfy the second-order linear ordinary differential equation

$$(1 - x^2)\frac{d^2y}{dx^2} - x\frac{dy}{dx} + n^2 y = 0. \tag{8}$$

The first few Chebyshev polynomials are given by

$$T_0(x) = 1, \quad T_1(x) = x, \quad T_2(x) = 2x^2 - 1, \quad T_3(x) = 4x^3 - 3x. \tag{9}$$

In the program we implement

$$T_n(x) := \frac{1}{2}(x + \sqrt{x^2 - 1})^n + \frac{1}{2}(x - \sqrt{x^2 - 1})^n \qquad (10)$$

as a procedure.

```
%cheby.red;

procedure T(n,x);
1/2*(x+sqrt(x^2-1))^n + 1/2*(x-sqrt(x^2-1))^n$

r1 := T(4,x);
r2 := T(5,x);
r3 := T(9,x);

%value of T(9,x) for x = 2;
x := 2;
r3;
clear x;

%r1 satisfies (7) with n = 4;
r4 := (1 - x*x)*df(r1,x,2) - x*df(r1,x) + 4*4*r1;

%r1, r2, r3 satisfy (2) with n = 5 and m = 4;
r5 := r3 + T(1,x) - 2*r2*r1;
```

The output is

```
R1 := 8*X**4 - 8*X**2 + 1$

R2 := X*(16*X**4 - 20*X**2 + 5)$

R3 := X*(256*X**8 - 576*X**6 + 432*X**4 - 120*X**2 + 9)$

70226$

R4 := 0$

R5 := 0$
```

## 45. Spherical Harmonics

Spherical harmonics are complex, single-valued functions defined on the surface of the unit sphere $S^2$. This means they are functions of two real-valued arguments $0 \leq \theta \leq \pi$, $0 \leq \phi < 2\pi$. They are denoted by $Y_{lm}$ with integer indices

$$l = 0, 1, 2 \ldots, \qquad m = -l, -l+1, \ldots, l-1, l. \tag{1}$$

Spherical harmonics play an important role in problems with spherical symmetry. The spherical harmonics $Y_{lm}$ are simultaneously eigenfunctions of the $z$-component $\hat{L}_z$ and the square $\hat{\mathbf{L}}^2$ of the total angular momentum operator $\hat{\mathbf{L}}$

$$\hat{L}_z Y_{lm} \equiv -i\frac{\partial}{\partial \phi} Y_{lm} = m Y_{lm} \tag{2}$$

$$\hat{\mathbf{L}}^2 Y_{lm} \equiv -\left(\frac{1}{\sin\theta}\frac{\partial}{\partial\theta}\left(\sin\theta\frac{\partial}{\partial\theta}\right) + \frac{1}{\sin^2\theta}\frac{\partial^2}{\partial\phi^2}\right) Y_{lm} = l(l+1) Y_{lm}. \tag{3}$$

As eigenfunctions of self-adjoint operators with different eigenvalues they form an orthogonal set of functions. The normalization condition is

$$\int_0^{2\pi} d\phi \int_0^{\pi} d\theta \sin\theta Y_{lm}^*(\theta, \phi) Y_{l'm'}(\theta, \phi) = \delta_{ll'}\delta_{mm'}. \tag{4}$$

Any well-behaved function defined on the surface of the unit sphere can be written as a (possibly) infinite linear combination of the spherical harmonics. The completeness relation is given by

$$\sum_{l=0}^{\infty}\sum_{m=-l}^{+l} Y_{lm}^*(\theta', \phi') Y_{lm}(\theta, \phi) = \delta(\phi - \phi')\delta(\cos\theta - \cos\theta'), \tag{5}$$

where $\delta$ denotes the delta function. Thus the spherical harmonics are an orthonormal basis of the Hilbert space $L_2(S^2)$. The program evaluates $Y_{lm}$ for a given $l$ and $m$.

```
%ylm.red;

procedure YY(l,m);    % see (46) for the formula;
begin
scalar a, b;
a := sin(w)^abs(m)/(2^l)*df((cos(w)^2-1)^l,cos(w),l+abs(m))
     /factorial(l);
if (m > 0) then a := (-1)^m*a;
b := sqrt((2*l+1)/(4*pi)*factorial(l-abs(m))
     /factorial(l+abs(m)));
return b*a*exp(i*m*phi);
end;

%Examples;
r1 := YY(0,0);   r2 := YY(1,1);
r3 := YY(1,-1);  r4 := YY(2,0);
r5 := YY(2,2);   r6 := YY(2,-1);

%We show that YY(2,2) satisfies (3) with l = 2 and m = 2;
r7 := 1/sin(w)*cos(w)*df(r5,w) + df(r5,w,2)
+ 1/(sin(w))**2*df(r5,phi,2) + 2*(2+1)*r5;
for all q let sin(q)**2+cos(q)**2=1;
r7;
```

The output is

```
R1 := (1/2)/SQRT(PI)$
R2 := ( - E**(I*PHI)*SQRT(3)*SIN(W))/(2*SQRT(PI)*SQRT(2))$
R3 := (SQRT(3)*SIN(W))/(2*SQRT(PI)*E**(I*PHI)*SQRT(2))$
R4 := (SQRT(5)*(3/2*COS(W)**2 - 1/2))/(2*SQRT(PI))$
R5 := (3*E**(2*I*PHI)*SQRT(5)*SIN(W)**2)/(4*SQRT(PI)*SQRT(6))$
R6 := (3*SQRT(5)*SIN(W)*COS(W))/(2*SQRT(PI)*E**(I*PHI)*SQRT(6))$

%We show that YY(2,2) satisfies (3) with l = 2 and m = 2;
r7 := 1/sin(w)*cos(w)*df(r5,w) + df(r5,w,2)
+ 1/(sin(w))**2*df(r5,phi,2) + 2*(2+1)*r5;

R7 := (3*E**(2*I*PHI)*SQRT(5)*(SIN(W)**2 + COS(W)**2 - 1))/
(SQRT(PI)*SQRT(6))$

0$
```

## 46. Clebsch-Gordon Series

Spherical harmonics are complex, single valued functions on the surface of the unit sphere, i.e. functions of two real-valued arguments $0 \leq \theta \leq \pi$, $0 \leq \phi < 2\pi$. They are denoted by $Y_{lm}$ and can be defined as

$$Y_{lm}(\theta, \phi) := \frac{(-1)^{l+m}}{2^l l!} \sqrt{\frac{2l+1}{4\pi} \frac{(l-m)!}{(l+m)!}} \sin^{|m|} \theta \frac{d^{l+|m|}}{d(\cos \theta)^{l+|m|}} (\sin \theta)^{2l} e^{im\phi}, \qquad (1)$$

where $l = 0, 1, 2, \ldots$ and $m = -l, -l+1, \ldots, +l$. The first few spherical harmonics are given by

$$Y_{00}(\theta, \phi) = \frac{1}{\sqrt{4\pi}}, \quad Y_{11}(\theta, \phi) = -\sqrt{\frac{3}{8\pi}} \sin \theta e^{i\phi}$$

$$Y_{10}(\theta, \phi) = -\sqrt{\frac{3}{4\pi}} \cos \theta, \quad Y_{1-1}(\theta, \phi) = \sqrt{\frac{3}{8\pi}} \sin \theta e^{-i\phi}. \qquad (2)$$

We have the relation

$$Y_{l,-m} = (-1)^m Y_{lm}^*. \qquad (3)$$

Any function in the Hilbert space $L_2(S^2)$ can be written as linear combinations of the spherical harmonics.

As a special example the direct product of two spherical harmonics may be expanded into the so-called Clebsch-Gordan series

$$Y_{l_1 m_1}(\theta, \phi) Y_{l_2 m_2}(\theta, \phi) = \sum_L \sum_M \sqrt{\frac{(2l_1+1)(2l_2+1)}{4\pi(2L+1)}} C^{L0}_{l_1 0 l_2 0} C^{LM}_{l_1 m_1 l_2 m_2} Y_{LM}(\theta, \phi). \qquad (4)$$

In the program we calculate the Clebsch Gordon series for $l_1 = 3$, $m_1 = -2$, $l_2 = 2$ and $m_2 = 1$. We apply the orthogonality of the spherical harmonics.

```
%cleb.red;

l1 := 3; m1 := -2; l2 := 2; m2 := 1;

if (abs(m1)>l1) or (abs(m2)>l2) or (l1<0) or (l2<0) then
<<write "indices out of range"; pause>>;

procedure YY(l,m);
begin
scalar a, b;
a := sin(w)^abs(m)/(2^l)*df((cos(w)^2-1)^l,cos(w),l+abs(m))/factorial(l);
if (m > 0) then a := (-1)^m*a;
b := sqrt((2*l+1)/(4*pi)*factorial(l-abs(m))/factorial(l+abs(m)));
return b*a*exp(i*m*phi);
end;

%P is the product of the spherical harmonics;
%We expand it into a series of spherical harmonics;

P := YY(l1,m1)*YY(l2,m2);

%Determination of the Clebsch-Gordon series;
%We use the orthogonality of the spherical harmonics;
%P is an eigenfunction of the z-component of the
%total angular momentum with eigenvalue m1+m2;

operator Y;
Clebsch := 0;

for l := abs(m1+m2) step 1 until l1+l2 do
    begin
    c := int(P*YY(l,m1+m2)*exp(-2*i*(m1+m2)*phi)*sin(w),w);
    let w = pi;  c2 := c;
    let w = 0;   c1 := c;
    clear w;
    Clebsch := Clebsch + (c2-c1)*Y(l,m1+m2);
    end;
Clebsch := 2*pi*Clebsch;
end;
```

The output is

```
P := ( - 15*SQRT(7)*SIN(W)**3*COS(W)**2)/(16*E**(I*PHI)*PI)$
CLEBSCH := (SQRT(PI)*SQRT(3)*(2*SQRT(770)*Y(5,-1) - 77*Y(3,-1) -
33*SQRT(14)*Y(1,-1)))/(462*PI)$
```

## 47. Hypergeometric Functions

The standard form of the hypergeometric differential equation is given by

$$x(1-x)\frac{d^2u}{dx^2} + (c - (a+b+1)x)\frac{du}{dx} - abu = 0. \tag{1}$$

A solution of (1) is given by

$$F(a,b,c;x) = 1 + \frac{ab}{c}\frac{x}{1!} + \frac{a(a+1)b(b+1)}{c(c+1)}\frac{x^2}{2!} + \frac{a(a+1)(a+2)b(b+1)(b+2)}{c(c+1)(c+2)}\frac{x^3}{3!} + \dots \tag{2}$$

If $a$, $b$, $c$ are real, then the series converges for $-1 < x < 1$ provided that $c - (a+b) > -1$.

If $c$, $a - b$ and $c - a - b$ are all nonintegers, the general solution valid for $|x| < 1$ is

$$u(x) = AF(a,b,c;x) + Bx^{1-c}F(a-c+1,b-c+1,2-c;x), \tag{3}$$

where $A$ and $B$ are the constants of integration. We find the following properties

$$F(a,b+1,c;x) - F(a,b,c;x) = \frac{ax}{c}F(a+1,b+1,c+1,x) \tag{4a}$$

$$F(a+1,b+1,c;x) - F(a,b,c;x) = \frac{abx}{c^2}F(a+1,b+1,c+1,x) \tag{4b}$$

$$\frac{d}{dx}F(a,b,c;x) = \frac{ab}{c}F(a+1,b+1,c+1;x). \tag{4c}$$

Many of the most important functions in analysis can be expressed by means of the hypergeometric series. We find

$$(1+x)^n = F(-n,b,b,-x) \tag{5a}$$

$$\ln(1+x) = xF(1,1,2;-x) \tag{5b}$$

$$\exp(x) = \lim_{\epsilon \to \infty} F\left(1,\epsilon,1;\frac{x}{\epsilon}\right). \tag{5c}$$

In the first program we insert the power series expansion

$$\sum_{n=0}^{8} f(n)x^n \tag{6}$$

into (1) and evaluate the expansion coefficients $f(0)$, $f(1)$, $f(2)$ and $f(3)$. We obtain the coefficients of series (2).

In the second program we insert

$$u(x) = (1-x)^4 \tag{7}$$

into (1) and determine the coefficients $a$, $b$, $c$.

The hypergeometric function arises in a large number of quantum mechanical problems (see 48).

```
%hypint.red;

operator f;

for n:=0 step 1 until 8 sum (x**n)*f(n);
u := ws;
res := x*(1-x)*df(u,x,2) + (c-(a+b+1)*x)*df(u,x,1) - a*b*u;

d0 := coeffn(res,x,0);
f(0) := 1;
list1 := solve({d0},{f(1)});
part(list1,1);
f(1) := part(ws,2);
d1 := coeffn(res,x,1);
list2 := solve({d1},{f(2)});
part(list2,1);
f(2) := part(ws,2);
d2 := coeffn(res,x,2);
list3 := solve({d2},{f(3)});
part(list3,1);
f(3) := part(ws,2);
d3 := coeffn(res,x,3);
list4 := solve({d3},{f(4)});
part(list4,1);
f(4) := part(ws,2);
```

The output is

```
U := F(8)*X**8 + F(7)*X**7 + F(6)*X**6 + F(5)*X**5 + F(4)*X**4 +
F(3)*X**3 + F(2)*X**2 + F(1)*X + F(0)$

RES :=  - F(8)*A*B*X**8 - 8*F(8)*A*X**8 - 8*F(8)*B*X**8 +
8*F(8)*C*X**7 - 64*F(8)*X**8 + 56*F(8)*X**7 - F(7)*A*B*X**7 -
7*F(7)*A*X**7 - 7*F(7)*B*X**7 + 7*F(7)*C*X**6 - 49*F(7)*X**7 +
42*F(7)*X**6 - F(6)*A*B*X**6 - 6*F(6)*A*X**6 - 6*F(6)*B*X**6 +
6*F(6)*C*X**5 - 36*F(6)*X**6 + 30*F(6)*X**5 - F(5)*A*B*X**5 -
5*F(5)*A*X**5 - 5*F(5)*B*X**5 + 5*F(5)*C*X**4 - 25*F(5)*X**5 +
20*F(5)*X**4 - F(4)*A*B*X**4 - 4*F(4)*A*X**4 - 4*F(4)*B*X**4 +
4*F(4)*C*X**3 - 16*F(4)*X**4 + 12*F(4)*X**3 - F(3)*A*B*X**3 -
3*F(3)*A*X**3 - 3*F(3)*B*X**3 + 3*F(3)*C*X**2 - 9*F(3)*X**3 +
6*F(3)*X**2 - F(2)*A*B*X**2 - 2*F(2)*A*X**2 - 2*F(2)*B*X**2 +
2*F(2)*C*X - 4*F(2)*X**2 + 2*F(2)*X - F(1)*A*B*X - F(1)*A*X -
F(1)*B*X + F(1)*C - F(1)*X - F(0)*A*B$

D0 := F(1)*C - F(0)*A*B$
F(0) := 1$

F(1) := (A*B)/C$
D1 := (2*F(2)*C**2 + 2*F(2)*C - A**2*B**2 - A**2*B - A*B**2 - A*B)/C$

F(2) := (A*B*(A*B + A + B + 1))/(2*C*(C + 1))$
D2 := (6*F(3)*C**3 + 18*F(3)*C**2 + 12*F(3)*C - A**3*B**3 -
3*A**3*B**2 - 2*A**3*B - 3*A**2*B**3 - 9*A**2*B**2 - 6*A**2*B -
2*A*B**3 - 6*A*B**2 - 4*A*B)/(2*C*(C + 1))$

F(3) := (A*B*(A**2*B**2 + 3*A**2*B + 2*A**2 + 3*A*B**2 + 9*A*B +
6*A + 2*B**2 + 6*B + 4))/(6*C*(C**2 + 3*C + 2))$
D3 := (24*F(4)*C**4 + 144*F(4)*C**3 + 264*F(4)*C**2 + 144*F(4)*C -
A**4*B**4 - 6*A**4*B**3 - 11*A**4*B**2 - 6*A**4*B - 6*A**3*B**4 -
36*A**3*B**3 - 66*A**3*B**2 - 36*A**3*B - 11*A**2*B**4 - 66*A**2*B**3 -
121*A**2*B**2 - 66*A**2*B - 6*A*B**4 - 36*A*B**3 - 66*A*B**2 - 36*A*B)/
(6*C*(C**2 + 3*C + 2))$

F(4) := (A*B*(A**3*B**3 + 6*A**3*B**2 + 11*A**3*B + 6*A**3 +
6*A**2*B**3 + 36*A**2*B**2 + 66*A**2*B + 36*A**2 + 11*A*B**3 +
66*A*B**2 + 121*A*B + 66*A + 6*B**3 + 36*B**2 + 66*B + 36))/
(24*C*(C**3 + 6*C**2 + 11*C + 6))$
```

In the next program we show that

$$u(x) = (1 - x)^4$$

is a solution of the hypergeometric differential equation if $a = -4$ and $c = b$ or $b = -4$ and $c = a$.

```
%hypint1.red;

depend u, x;
u := (1 - x)**4;

res := x*(1-x)*df(u,x,2) + (c-(a+b+1)*x)*df(u,x) - a*b*u;

c0 := coeffn(res,x,0);
c1 := coeffn(res,x,1);
c2 := coeffn(res,x,2);
c3 := coeffn(res,x,3);
c4 := coeffn(res,x,4);

list := solve({c0=0,c1=0,c2=0,c3=0,c4=0},{a,b,c});
```

The output is

```
u := x**4 - 4*x**3 + 6*x**2 - 4*x + 1$

res := - a*b*x**4 + 4*a*b*x**3 - 6*a*b*x**2 + 4*a*b*x -
a*b - 4*a*x**4 + 12*a*x**3 - 12*a*x**2 + 4*a*x - 4*b*x**4 +
12*b*x**3 - 12*b*x**2 + 4*b*x + 4*c*x**3 - 12*c*x**2 + 12*c*x -
4*c - 16*x**4 + 48*x**3 - 48*x**2 + 16*x$

c0 :=  - a*b - 4*c$
c1 := 4*(a*b + a + b + 3*c + 4)$
c2 := 6*( - a*b - 2*a - 2*b - 2*c - 8)$
c3 := 4*(a*b + 3*a + 3*b + c + 12)$
c4 :=  - a*b - 4*a - 4*b - 16$

list := {{a=-4,b=arbcomplex(2),c=b},{b=-4,a=arbcomplex(1),c=a}}$
```

## 48. Eigenvalue Problem and Hypergeometric Differential Equation

A number of one-dimensional eigenvalue problems

$$-\frac{\hbar^2}{2m}\frac{d^2u}{dx^2} + V(x)u = Eu, \quad \text{or} \quad \frac{d^2u}{dx^2} + \frac{2m}{\hbar^2}(E - V(x))u = 0 \tag{1}$$

can be cast into the hypergeometric differential equation (see 47)

$$y(1-y)\frac{d^2v}{dy^2} + (c - (a+b+1))\frac{dv}{dy} - abv = 0. \tag{2}$$

It should be mentioned that some radial symmetric eigenvalue problems can also be cast into the hypergeometric differential equation.

A transformation of the dependent and independent variables is applied to obtain the hypergeometric differential equation. Let

$$y(x) = Q(x), \quad v(y(x)) = P(x)u(x) \tag{3}$$

be the invertible transformation. Applying the chain rule we find

$$\frac{dv}{dx} = \frac{dv}{dy}\frac{dy}{dx} = \frac{dP}{dx}u + P\frac{du}{dx}, \tag{4}$$

where $dy/dx = dQ/dx$. Therefore

$$\frac{du}{dx} = \frac{1}{P}\frac{dQ}{dx}\frac{dv}{dy} - \frac{1}{P^2}\frac{dP}{dx}v, \tag{5}$$

where we used the fact that $u = v/P$. The second derivative of $v$ with respect to $x$ yields

$$\frac{d^2v}{dy^2}\frac{dy}{dx}\frac{dQ}{dx} + \frac{dv}{dy}\frac{d^2Q}{dx^2} = \frac{d^2P}{dx^2}u + 2\frac{dP}{dx}\frac{du}{dx} + P\frac{d^2u}{dx^2}. \tag{6}$$

Therefore

$$\frac{d^2u}{dx^2} = \frac{1}{P}\left(\frac{dQ}{dx}\right)^2\frac{d^2v}{dy^2} + \left(\frac{1}{P}\frac{d^2Q}{dx^2} - \frac{2}{P^2}\frac{dP}{dx}\frac{dQ}{dx}\right)\frac{dv}{dy} + \left(-\frac{1}{P^2}\frac{d^2P}{dx^2} + \frac{2}{P^3}\left(\frac{dP}{dx}\right)^2\right)v. \quad (7)$$

The eigenvalue equation can be written in the form

$$\frac{d^2u}{dx^2} + B(x)u = 0. \tag{8}$$

Inserting $u = v/P$ and (7) into (8) and multiplying by $P$ yields

$$\left(\frac{dQ}{dx}\right)^2\frac{d^2v}{dy^2} + \left(\frac{d^2Q}{dx^2} - \frac{2}{P}\frac{dP}{dx}\frac{dQ}{dx}\right)\frac{dv}{dy} + \left(\frac{2}{P^2}\left(\frac{dP}{dx}\right)^2 - \frac{1}{P}\frac{d^2P}{dx^2} + B(x)\right)v = 0, \quad (9)$$

where the invertible transformation (3) has to be inserted to eliminate $x$.

In the program we consider the case

$$\frac{d^2u}{dx^2} + \frac{2m}{\hbar^2}\left(E + \frac{V_0}{\cosh^2(x/a)}\right)u = 0. \tag{10}$$

Therefore

$$B(x) = \frac{2m}{\hbar^2}\left(E + \frac{V_0}{\cosh^2(x/a)}\right). \tag{11}$$

Transformation (3) is given by

$$y(x) = Q(x) = -\sinh^2(x/a), \qquad P(x) = \left(\cosh\left(\frac{x}{a}\right)\right)^{2\lambda}, \tag{12}$$

where

$$\lambda := \frac{1}{4} \left( \sqrt{\frac{8mV_0 a^2}{\hbar^2} + 1} - 1 \right).$$ (13)

We obtain

$$y(1-y)\frac{d^2v}{dy^2} + \left( \frac{1}{2} - (1-2\lambda)y \right) \frac{dv}{dy} - (\lambda^2 + k^2)v = 0,$$ (14)

where

$$k = \sqrt{-\frac{mEa^2}{2\hbar^2}}.$$ (15)

```
%trans.red;

depend Q, x;
depend P, x;
depend B, x;
depend v, y;

res1 := df(Q,x)*df(Q,x)*df(v,y,2) + (df(Q,x,2) -
(2/P)*df(P,x)*df(Q,x))*df(v,y) +
(2/(P*P)*df(P,x)*df(P,x)-(1/P)*df(P,x,2)+B)*v;

P := (cosh(x/a))**(2*lam);
Q := -sinh(x/a)*sinh(x/a);
B := (2*m/(hb*hb))*(E + V0/(cosh(x/a)*cosh(x/a)));

res1;

for all c let cosh(c) = sqrt(1 + sinh(c)**2);
res1;

lam := 1/4*(sqrt(8*m*V0*a*a/(hb*hb) + 1) - 1);
res1;
res2 := sub(sinh(x/a)=sqrt(-z),res1);
```

The output is

```
res1;

(2*(2*DF(V,Y,2)*COSH(X/A)**4*SINH(X/A)**2*HB**2 -
DF(V,Y)*COSH(X/A)**4*HB**2 +
4*DF(V,Y)*COSH(X/A)**2*SINH(X/A)**2*LAM*HB**2 -
DF(V,Y)*COSH(X/A)**2*SINH(X/A)**2*HB**2 +
COSH(X/A)**2*A**2*E*M*V - COSH(X/A)**2*V*LAM*HB**2 +
2*SINH(X/A)**2*V*LAM**2*HB**2 + SINH(X/A)**2*V*LAM*HB**2 +
A**2*M*V*V0))/(COSH(X/A)**2*A**2*HB**2)$


res1;

(2*(2*DF(V,Y,2)*SINH(X/A)**10*HB**2 +
8*DF(V,Y,2)*SINH(X/A)**8*HB**2 +
12*DF(V,Y,2)*SINH(X/A)**6*HB**2 +
8*DF(V,Y,2)*SINH(X/A)**4*HB**2 +
2*DF(V,Y,2)*SINH(X/A)**2*HB**2 +
4*DF(V,Y)*SINH(X/A)**8*LAM*HB**2 -
2*DF(V,Y)*SINH(X/A)**8*HB**2 +
12*DF(V,Y)*SINH(X/A)**6*LAM*HB**2 -
7*DF(V,Y)*SINH(X/A)**6*HB**2 +
12*DF(V,Y)*SINH(X/A)**4*LAM*HB**2 -
9*DF(V,Y)*SINH(X/A)**4*HB**2 +
4*DF(V,Y)*SINH(X/A)**2*LAM*HB**2 -
5*DF(V,Y)*SINH(X/A)**2*HB**2 -
DF(V,Y)*HB**2 +SINH(X/A)**6*A**2*E*M*V +
2*SINH(X/A)**6*V*LAM**2*HB**2 +
3*SINH(X/A)**4*A**2*E*M*V + SINH(X/A)**4*A**2*M*V*V0 +
4*SINH(X/A)**4*V*LAM**2*HB**2 - SINH(X/A)**4*V*LAM*HB**2 +
3*SINH(X/A)**2*A**2*E*M*V + 2*SINH(X/A)**2*A**2*M*V*V0 +
2*SINH(X/A)**2*V*LAM**2*HB**2 - 2*SINH(X/A)**2*V*LAM*HB**2 +
A**2*E*M*V + A**2*M*V*V0 - V*LAM*HB**2))/
(A**2*HB**2*(SINH(X/A)**6 + 3*SINH(X/A)**4 + 3*SINH(X/A)**2 + 1))$

RES2 := ( - 4*SQRT(8*A**2*M*V0 + HB**2)*DF(V,Y)*Z*HB -
SQRT(8*A**2*M*V0 + HB**2)*V*HB + 8*DF(V,Y,2)*Z**2*HB**2 -
8*DF(V,Y,2)*Z*HB**2 + 12*DF(V,Y)*Z*HB**2 - 4*DF(V,Y)*HB**2 +
4*A**2*E*M*V + 4*A**2*M*V*V0 + V*HB**2)/(2*A**2*HB**2)$
```

## 49. Gamma Matrices and Spin Matrices

The Pauli spin matrices $\sigma_x$, $\sigma_y$, $\sigma_z$ and the gamma matrices $\gamma_1$, $\gamma_2$, $\gamma_3$, $\gamma_4$ play a central role in describing the electron (or more general Fermi particles). The Pauli spin matrices are given by

$$\sigma_x := \begin{pmatrix} 0 & 1 \\ 1 & 0 \end{pmatrix}, \quad \sigma_y := \begin{pmatrix} 0 & -i \\ i & 0 \end{pmatrix}, \quad \sigma_z := \begin{pmatrix} 1 & 0 \\ 0 & -1 \end{pmatrix}. \tag{1}$$

The gamma matrices are given by

$$\gamma_1 := \begin{pmatrix} 0 & 0 & 0 & -i \\ 0 & 0 & -i & 0 \\ 0 & i & 0 & 0 \\ i & 0 & 0 & 0 \end{pmatrix}, \quad \gamma_2 := \begin{pmatrix} 0 & 0 & 0 & -1 \\ 0 & 0 & 1 & 0 \\ 0 & 1 & 0 & 0 \\ -1 & 0 & 0 & 0 \end{pmatrix}$$

$$\gamma_3 := \begin{pmatrix} 0 & 0 & -i & 0 \\ 0 & 0 & 0 & i \\ i & 0 & 0 & 0 \\ 0 & -i & 0 & 0 \end{pmatrix}, \quad \gamma_4 := \begin{pmatrix} 1 & 0 & 0 & 0 \\ 0 & 1 & 0 & 0 \\ 0 & 0 & -1 & 0 \\ 0 & 0 & 0 & -1 \end{pmatrix}. \tag{2}$$

Let $I_2$ be the $2 \times 2$ unit matrix. Then the gamma matrices can be expressed as

$$\gamma_1 = \sigma_y \otimes \sigma_x, \quad \gamma_2 = \sigma_y \otimes \sigma_y, \quad \gamma_3 = \sigma_y \otimes \sigma_z, \quad \gamma_4 = \sigma_z \otimes I_2. \tag{3}$$

In the program we implement the Kronecker product as a procedure and then calculate the gamma matrices from the Pauli spin matrices and the unit matrix by applying (3).

```
%gamma.red;

procedure Kron(A,B);
begin
n := 2; m := n*n;

operator A$  matrix AA(n,n)$
for i:=1:n do
for j:=1:n do
AA(i,j):=A(i,j)$

operator B$  matrix BB(n,n)$
for i:=1:n do
for j:=1:n do
BB(i,j):=B(i,j)$

operator C$  matrix CC(m,m);
c1 := 0; c2 := 0;
for r:=1:n do
for s:=1:n do
begin
for i:=1:n do
for j:=1:n do
begin
c1 := n*(r-1); c2 := n*(s-1);
CC(i+c1,j+c2) := AA(r,s)*BB(i,j);
end;  end;
return CC;
end;    % end procedure Kron;

operator sx, sy, sz, ID;
sx(1,1) := 0; sx(1,2) := 1; sx(2,1) := 1; sx(2,2) := 0;
sy(1,1) := 0; sy(1,2) := -i; sy(2,1) := i; sy(2,2) := 0;
sz(1,1) := 1; sz(1,2) := 0; sz(2,1) := 0; sz(2,2) := -1;
ID(1,1) := 1; ID(1,2) := 0; ID(2,1) := 0; ID(2,2) := 1;

g1 := Kron(sy,sx); g2 := Kron(sy,sy);
g3 := Kron(sy,sz); g4 := Kron(sz,ID);
```

The output is

```
g1 := mat((0,0,0, - i),(0,0, - i,0),(0,i,0,0),(i,0,0,0))$
g2 := mat((0,0,0,-1),(0,0,1,0),(0,1,0,0),(-1,0,0,0))$
g3 := mat((0,0, - i,0),(0,0,0,i),(i,0,0,0),(0, - i,0,0))$
g4 := mat((1,0,0,0),(0,1,0,0),(0,0,-1,0),(0,0,0,-1))$
```

## 50. Fourier Transform

Let $L_1(\mathcal{R})$ be the Lebesgue space of integrable functions over $\mathcal{R}$. Then the Fourier transform is defined by

$$\hat{f}(k) := \int\limits_{-\infty}^{\infty} e^{ikx} f(x) dx. \tag{1}$$

We find that $\hat{f} \in L_1(\mathcal{R})$. The inverse transformation is given by

$$f(x) = \frac{1}{2\pi} \int\limits_{-\infty}^{\infty} e^{-ikx} \hat{f}(k) dk. \tag{2}$$

We have

$$\left| \int\limits_{-\infty}^{\infty} e^{ikx} f(x) dx \right| \leq \int\limits_{-\infty}^{\infty} |f(x)| dx \tag{3}$$

and

$$\int\limits_{-\infty}^{\infty} f(x) g^*(x) dx = \frac{1}{2\pi} \int\limits_{-\infty}^{\infty} \hat{f}(k) \hat{g}^*(k) dk \qquad \text{Parseval's equation.} \tag{4}$$

In the program we consider the following examples:

Example 1: We consider

$$f_a(x) = \begin{cases} \frac{1}{2a} & \text{for } |x| < a \\ 0 & \text{for } |x| > a \end{cases}, \tag{5}$$

where $a > 0$. Obviously $f \in L_1(\mathcal{R})$. We find that

$$\int_{-\infty}^{\infty} f_a(x)dx = 1. \tag{6}$$

For the Fourier transform we find

$$\hat{f}_a(k) = \frac{\sin(ak)}{ak}. \tag{7}$$

Example 2: We consider

$$g_a(x) = \frac{a}{2}\exp(-a|x|), \tag{8}$$

where $a > 0$. Then

$$\int_{-\infty}^{\infty} g_a(x)dx = 1. \tag{9}$$

For the Fourier transform we find

$$\hat{g}_a(k) = \frac{a^2}{a^2 + k^2}. \tag{10}$$

Notice that $g_a$ is not differentiable at $x = 0$, whereas $\hat{g}_a$ is differentiable everywhere.

Example 3: We consider

$$h(x) = \begin{cases} 2x & \text{for} & 0 \le x \le 1/2 \\ 2(1-x) & \text{for} & 1/2 \le x \le 1 \\ 0 & \text{otherwise} \end{cases} \tag{11}$$

We have

144

$$\int\limits_{-\infty}^{\infty} h(x)dx = \frac{1}{2}. \tag{12}$$

The Fourier transform is given by

$$\hat{h}(k) = \frac{2(2i\sin(k/2) - i\sin(k) + 2\cos(k/2) - \cos(k) - 1)}{k^2}. \tag{13}$$

In the program we set $\hat{f}_a \to fah$, $\hat{g}_a \to gah$ and $\hat{h} \to hh$.

```
%ft1.red;

depend ef, x;
for all y let exp(i*y) = cos(y) + i*sin(y);
ef := exp(i*k*x);

%example 1;
depend fa, x;
fa := ef/(2*a);
fah := sub(x=a,int(fa,x))-sub(x=-a,int(fa,x));

%example 2;
depend ga1, x;
depend ga2, x;
ga1 := a*exp(a*x)/2;
ga2 := a*exp(-a*x)/2;
int2 := sub(x=0,int(ga1,x)) - sub(x=0,int(ga2,x));
gah := sub(x=0,int(ga1*ef,x)) - sub(x=0,int(ga2*ef,x));

%example 3;
depend h1, x;
h1 := 2*x;
depend h2, x;
h2 := 2*(1-x);
ht1 := sub(x=1/2,int(h1*ef,x))-sub(x=0,int(h1*ef,x));
ht2 := sub(x=1,int(h2*ef,x))-sub(x=1/2,int(h2*ef,x));
hh := ht1 + ht2;
```

The output is

```
%example 1;
fah := sin(a*k)/(a*k)$

%example 2;
int2 := 1$
gah := a**2/(a**2 + k**2)$

%example 3;
hh := (2*(2*sin(k/2)*i-sin(k)*i+2*cos(k/2)-cos(k)-1))/k**2$
```

*Remark:* Since $a > 0$ in example 1, we find that

$$\lim_{k \to 0} \frac{\sin(ak)}{ak} = 1.$$

Using the rule of L'Hospital this result can also be obtained from the following REDUCE program

```
%lhos.red

depend fah, k;
fah := sin(a*k)/(a*k);

N := num(fah);
D := den(fah);

ND := df(N,k);
DD := df(D,k);

fah0 := sub(k=0,ND)/sub(k=0,DD);
```

## 51. Discrete Fourier Transform

The discrete Fourier transform is an approximation of the continuous Fourier transformation (see 50). The discrete Fourier transform is used when a set of sample function values, $x(j)$, are available at equally spaced time intervals numbered $j = 0, 1, \ldots, N-1$. The discrete Fourier transform maps the given function values into the sum of a discrete number of sine waves whose frequencies are numbered $k = 0, 1, \ldots, N-1$, and whose amplitudes are given by

$$\hat{x}(k) = \frac{1}{N} \sum_{l=0}^{N-1} x(l) \exp\left(-i2\pi k \frac{l}{N}\right). \tag{1}$$

Equation (1) can also be written in terms of a real and imaginary part

$$\hat{x}(k) = \frac{1}{N} \sum_{l=0}^{N-1} x(l) \cos\left(2\pi k \frac{l}{N}\right) - \frac{i}{N} \sum_{l=0}^{N-1} x(l) \sin\left(2\pi k \frac{l}{N}\right). \tag{2}$$

The inverse transformation is given by

$$x(l) = \sum_{k=0}^{N-1} \hat{x}(k) \exp\left(i2\pi k \frac{l}{N}\right). \tag{3}$$

To find (3) we use the fact that

$$\sum_{k=0}^{N-1} \exp\left(i2\pi k \frac{n-m}{N}\right) = N\delta_{nm}, \tag{4}$$

where $n$ and $m$ are integers. In the program the input data are

$$x(k) = \cos(2\pi k/N), \qquad N = 8, \quad k = 0, 1, 2, \ldots, N-1. \tag{5}$$

The discrete Fourier transform is denoted by $xh(j)$ in the program. We also evaluate the inverse transformation.

```
%dfour.red;

n := 8;
array x(n);
array xh(n);  % Fourier transform;

% data input;
for k:=0:(n-1) do
x(k) := cos(2*pi*k/n);

for j:=0:(n-1) do
<<cossum := 0; sinsum := 0;
for k := 0:(n-1) do
<<ang := 2*pi*k*j/n;
cossum := cossum + x(k)*cos(ang);
sinsum := sinsum - x(k)*sin(ang) >>;
xh(j) := (cossum + i*sinsum)/n >>;

for j:=0:(n-1) do
write " xh(",j,")= ", xh(j);

% inverse transformation;
for j:=0:(n-1) do
<<cossum := 0; sinsum := 0;
for k := 0:(n-1) do
<<ang := 2*pi*k*j/n;
cossum := cossum + xh(k)*cos(ang);
sinsum := sinsum + xh(k)*sin(ang) >>;
x(j) := cossum + i*sinsum >>;

for j:=0:(n-1) do
write " x(",j,")= ", x(j);
```

The output is

```
xh(0)= 0$ xh(1)= 1/2$ xh(2)= 0$ xh(3)= 0$
xh(4)= 0$ xh(5)= 0$ xh(6)= 0$ xh(7)= 1/2$

x(0)= 1$ x(1)= sqrt(2)/2$ x(2)= 0$ x(3)= (-sqrt(2))/2$
x(4)= -1$ x(5)= (-sqrt(2))/2$ x(6)= 0$ x(7)= sqrt(2)/2$
```

## 52. Fourier Expansion

Consider a Hilbert space $\mathcal{H}$. Let

$$\mathcal{B} := \{\phi_j \ : \ j \in I, \ I \text{ countable index set}\} \tag{1}$$

be an (orthonormal) basis in the Hilbert space $\mathcal{H}$. Then every $f \in \mathcal{H}$ can be expressed as

$$f = \sum_{j \in I} a_j \phi_j, \tag{2}$$

where the (complex) expansion coefficients $a_j$ are given by

$$a_j := \langle f, \phi_j \rangle. \tag{3}$$

Here $\langle \, , \, \rangle$ denotes the scalar product.

As an example we consider the Hilbert space $L_2(0,1)$. A basis is given by

$$\mathcal{B} := \{e^{2\pi i k x} \ : \ k \in \mathcal{Z}\}. \tag{4}$$

The scalar product in $L_2(0,1)$ is given by

$$\langle f, g \rangle := \int\limits_0^1 f(x)g(x)^* dx. \tag{5}$$

The integration is performed in the Lebesgue sense. Let $f$ be given by

$$f(x) = \begin{cases} 2x & \text{for} \ \ 0 \le x \le 1/2 \\ 2(1-x) & \text{for} \ \ 1/2 \le x \le 1 \end{cases} \tag{6}$$

Thus the expansion coefficients are given by

$$a_k = \int\limits_0^1 f(x)e^{-2\pi i k x} dx = \int\limits_0^{1/2} 2x e^{-2\pi i k x} dx + \int\limits_{1/2}^1 2(1-x)e^{-2\pi i k x} dx. \tag{7}$$

In the program we evaluate $a_k$. The case $k = 0$ we evaluate separately. We can also find $a_0$ from $a_k$ by applying the L'Hospital rule.

```
%fexpa.red;

depend f1, x;
depend f2, x;

f1 := 2*x;
f2 := 2*(1-x);

r1 := sub(x=1/2,int(f1*exp(-2*i*pi*k*x),x))
-sub(x=0,int(f1*exp(-2*i*pi*k*x),x)));

r2 := sub(x=1,int(f2*exp(-2*i*pi*k*x),x))
-sub(x=1/2,int(f2*exp(-2*i*pi*k*x),x)));

ak := r1 + r2;

a0 := sub(x=1/2,int(f1,x)) - sub(x=0,int(f1,x))
+ sub(x=1,int(f2,x)) - sub(x=1/2,int(f2,x));
```

The output is

```
r1:=(-e**(i*k*pi)+i*k*pi+1)/(2*e**(i*k*pi)*k**2*pi**2)$

r2:=(-e**(i*k*pi)*i*k*pi+e**(i*k*pi)-1)/(2*e**(2*i*k*pi)*k**2*pi**2)$

ak:=(-e**(2*i*k*pi)+2*e**(i*k*pi)-1)/(2*e**(2*i*k*pi)*k**2*pi**2)$

a0 := 1/2$
```

## 53. Group Theory

Finite group theory plays an important role in quantum mechanics. First we introduce the definition of a group and describe some properties of a group.

A group is an abstract mathematical entity which expresses the intuitive concept of symmetry.

*Definition:* A *group* $G$ is a set of objects $\{g, h, k, \ldots\}$ (not necessarily countable) together with a binary operation which associates with any ordered pair of elements $g, h$ in $G$ a third element $gh$. The binary operation (called group multiplication) is subject to the following requirements:

(1) There exists an element $e$ in $G$ called the identity element such that $ge = eg = g$ for all $g \in G$.
(2) For every $g \in G$ there exists in $G$ an inverse element $g^{-1}$ such that

$$gg^{-1} = g^{-1}g = e.$$

(3) Associative law: The identity

$$(gh)k = g(hk)$$

is satisfied for all $g, h, k \in G$.

Thus, any set, together with a binary operation, which satisfies conditions (1)-(3) is called a group. If $gh = hg$ we say that the elements $g$ and $h$ commute. If all elements of $G$ commute then $G$ is a commutative or abelian group. If $G$ has a finite number of elements it has finite order $n(G)$, where $n(G)$ is the number of elements. Otherwise, $G$ has infinite order.

A subgroup $H$ of $G$ is a subset which is itself a group under the group multiplication defined in $G$. The subgroups $G$ and $\{e\}$ are called improper subgroups of $G$. All other subgroups are proper.

Example: Let $G = \{+1, -1\}$ and the binary operation be multiplication. Then $G$ is an abelian group.

Example: Let $GL(n, \mathcal{R})$ be the set of all invertible $n \times n$ matrices over $\mathcal{R}$. Then $GL(n, \mathcal{R})$ forms a group under matrix multiplication.

Example: All $n \times n$ unitary matrices form a group under matrix multiplication. A matrix $U$ is called unitary if $U^* = U^{-1}$.

A way to partition $G$ is by means of *conjugacy classes*.

*Definition:* A group element $h$ is said to be conjugate to the group element $k, h \sim k$, if there exists a $g \in G$ such that

$$k = ghg^{-1}.$$

It is easy to show that conjugacy is an equivalence relation, i.e., (1) $h \sim h$ (reflexive), (2) $h \sim k$ implies $k \sim h$ (symmetric), and (3) $h \sim k, k \sim j$ implies $h \sim j$ (transitive). Thus, the elements of $G$ can be divided into *conjugacy classes* of mutually conjugate elements. The class containing $e$ consists of just one element since

$$geg^{-1} = e$$

for all $g \in G$. Different conjugacy classes do not necessarily contain the same number of elements.

Example: Let $G$ be an abelian group. Then each conjugacy class consists of one group element each, since
$$ghg^{-1} = h, \quad \text{for all} \quad g \in G.$$

In the program we consider the permutation group $S_3$. We give the composition of the group elements. Then we evaluate the inverse of each group element. Finally we determine the conjugacy classes. The group consists of six elements which we denote by $a(0), a(1), \ldots, a(5)$. The neutral (identity) element is denoted by $a(0)$. Thus we have

$$a(0) * a(j) = a(j) * a(0) = a(j)$$

for $j = 0, 1, \ldots, 5$. The group is nonabelian.

```
%gr1.red;

operator a, g, res, test, cl1, cl2;
noncom a, g, res, test, cl1, cl2;

% a(0) is the neutral element;
for all j let a(j)*a(0) = a(j);
for all j let a(0)*a(j) = a(j);

let a(1)*a(1) = a(0);
let a(1)*a(2) = a(3); let a(2)*a(1) = a(4);
let a(1)*a(3) = a(2); let a(3)*a(1) = a(5);
let a(1)*a(4) = a(5); let a(4)*a(1) = a(2);
let a(1)*a(5) = a(4); let a(5)*a(1) = a(3);
let a(2)*a(2) = a(0);
let a(2)*a(3) = a(5); let a(3)*a(2) = a(1);
let a(2)*a(4) = a(1); let a(4)*a(2) = a(5);
let a(2)*a(5) = a(3); let a(5)*a(2) = a(4);
let a(3)*a(3) = a(4);
let a(3)*a(4) = a(0); let a(4)*a(3) = a(0);
let a(3)*a(5) = a(2); let a(5)*a(3) = a(1);
let a(4)*a(4) = a(3);
let a(4)*a(5) = a(1); let a(5)*a(4) = a(2);
let a(5)*a(5) = a(0);
res := a(0)*a(1)*a(2)*a(3)*a(4)*a(5);

% find the inverse;
g(0) := a(0);
for j:=0:5 do
begin
for k:=0:5 do
begin
test := a(j)*a(k);
if test = a(0) then g(j) := a(k);
end;
end;

for j:=0:5 do
write g(j);

% conjugacy class of the group element a(1);
for j:= 0:5 do
cl1(j) := a(j)*a(1)*g(j);
```

```
for j:=0:5 do
write cl1(j);

% conjugacy class of the group element a(3);
for j:=0:5 do
cl2(j) := a(j)*a(3)*g(j);

for j:=0:5 do
write cl2(j);
```

The output is

```
res := a(2)
```

```
for j:=0:5 do write g(j);
a(0)
a(1)
a(2)
a(4)
a(3)
a(5)
```

```
for j:=0:5 do write cl1(j);
a(1)
a(1)
a(5)
a(2)
a(5)
a(2)
```

```
for j:=0:5 do write cl2(j);
a(3)
a(4)
a(4)
a(3)
a(3)
a(4)
```

## 54. Quantum Groups

The quantum Yang-Baxter equation plays an important role both in physics and mathematics. With solutions of the quantum Yang-Baxter equation one can construct exactly solvable models and find their eigenvalues and eigenstates. On the other hand, any solution of the quantum Yang-Baxter equation can be generally used to find the new quasi-triangular Hopf algebra. Many multiparameter solutions ($4 \times 4$ matrices) of the quantum Yang-Baxter equation have been found. Corresponding to the case of the standard one-parameter $R$ matrix, the algebras related to the standard two-parameter $R$ matrix have also been discussed.

Consider the $2 \times 2$ matrix

$$T = \begin{pmatrix} a & b \\ c & d \end{pmatrix}, \tag{1}$$

where $a$, $b$, $c$ and $d$ are noncommutative linear operators. We may consider $a$, $b$, $c$ and $d$ as $n \times n$ matrices over the complex or real numbers. Let $I$ be the $2 \times 2$ unit matrix

$$I = \begin{pmatrix} 1 & 0 \\ 0 & 1 \end{pmatrix}, \tag{2}$$

where 1 is the unit operator (unit matrix). Now we define

$$T_1 := T \otimes I, \qquad T_2 = I \otimes T, \tag{3}$$

where $\otimes$ denotes the Kronecker product. Thus $T_1$ and $T_2$ are operator valued $4 \times 4$ matrices. Thus we find

$$T_1 = T \otimes I = \begin{pmatrix} a & 0 & b & 0 \\ 0 & a & 0 & b \\ c & 0 & d & 0 \\ 0 & c & 0 & d \end{pmatrix}, \qquad T_2 = I \otimes T = \begin{pmatrix} a & b & 0 & 0 \\ c & d & 0 & 0 \\ 0 & 0 & a & b \\ 0 & 0 & c & d \end{pmatrix}. \tag{4}$$

Consider now the 4 × 4 matrix

$$R_q := \begin{pmatrix} 1 & 0 & 0 & 0 \\ 0 & -1 & 0 & 0 \\ 0 & 1+q & q & 0 \\ 0 & 0 & 0 & 1 \end{pmatrix}. \tag{5}$$

The algebra related to this quantum matrix is governed by the Yang-Baxter equation

$$R_q T_1 T_2 = T_2 T_1 R_q. \tag{6}$$

Equation (6) gives rise to the relations of the algebra elements $a$, $b$, $c$ and $d$

$$ab = q^{-1}ba, \quad dc = qcd, \quad bc = -qcb,$$

$$bd = -db, \quad ac = -ca, \quad [a, d] = (1 + q^{-1})bc \tag{7}$$

where all the commutative relations have been omitted. This quantum matrix $T$ can be considered as a linear transformation of plane $\mathcal{A}_q(2)$ with coordinates $(x, \xi)$ satisfying $x\xi = -\xi x$. The coordinate transformations deduced by $T$

$$\begin{pmatrix} x' \\ \xi' \end{pmatrix} = \begin{pmatrix} a & b \\ c & d \end{pmatrix} \begin{pmatrix} x \\ \xi \end{pmatrix} \tag{8}$$

keep the relation $x'\xi' = -\xi'x'$. As there is no nilpotential element in quantum matrix $T$, there exists no constraint on the coordinates $x$ and $\xi$. This quantum plane is the same as the one related to $GL_q(2)$. From the algebraic relations (7) we can define an element of the algebra,

$$\delta(T) = \delta = ad - q^{-1}bc. \tag{9}$$

$\delta$ satisfies the following relations

$$[a, \delta] = 0, \quad [d, \delta] = 0$$

$$\{c, \delta\}_q \equiv qc\delta + \delta c = 0, \quad \{\delta, b\}_q \equiv q\delta b + b\delta = 0. \tag{10}$$

The element $\delta$ commutes only with $a$ and $b$ and hence is not the centre of the algebra.

In the program we implement the Kronecker product. Then we evaluate the matrices $T_1$ and $T_2$. Finally we evaluate condition (6) and thus find the commutation relations (7).

```
%qg.red;

procedure Kron(A,B);
begin
n := 2; m := n*n;
operator A$
matrix AA(n,n)$
for i:=1:n do
for j:=1:n do
AA(i,j):=A(i,j)$
operator B$
matrix BB(n,n)$
for i:=1:n do
for j:=1:n do
BB(i,j):=B(i,j)$

operator C$
matrix CC(m,m);
c1 := 0; c2 := 0;
for r:=1:n do
for s:=1:n do
begin
for i:=1:n do
for j:=1:n do
begin
c1 := n*(r-1); c2 := n*(s-1);
CC(i+c1,j+c2) := AA(r,s)*BB(i,j);
end;
end;
return CC;
end;

operator a, b, c, d, id;
noncom a, b, c, d;

operator T;
matrix T1(4,4);
matrix T2(4,4);
matrix R(4,4);

T(1,1) := a(j); T(1,2) := b(j);
T(2,1) := c(j); T(2,2) := d(j);

ID(1,1) := 1; ID(1,2) := 0; ID(2,1) := 0; ID(2,2) := 1;
```

```
T1 := Kron(T,ID);
T2 := Kron(ID,T);

R(1,1) := 1; R(1,2) := 0;   R(1,3) := 0; R(1,4) := 0;
R(2,1) := 0; R(2,2) := -1;  R(2,3) := 0; R(2,4) := 0;
R(3,1) := 0; R(3,2) := 1+q; R(3,3) := q; R(3,4) := 0;
R(4,1) := 0; R(4,2) := 0;   R(4,3) := 0; R(4,4) := 1;

matrix RES(4,4);

RES := R*T1*T2 - T2*T1*R;
```

The output is

```
t1 := mat((a(j),0,b(j),0),(0,a(j),0,b(j)),(c(j),0,d(j),0),
(0,c(j),0,d(j)))$

t2 := mat((a(j),b(j),0,0),(c(j),d(j),0,0),(0,0,a(j),b(j)),
(0,0,c(j),d(j)))$

res := mat((0, - a(j)*b(j)*q + b(j)*a(j),
- a(j)*b(j)*q + b(j)*a(j),0),
( - (a(j)*c(j) + c(j)*a(j))),
- a(j)*d(j) - c(j)*b(j)*q - c(j)*b(j) + d(j)*a(j),
- (b(j)*c(j) + c(j)*b(j)*q), - (b(j)*d(j) + d(j)*b(j))),
(q*(a(j)*c(j) + c(j)*a(j)),b(j)*c(j) + c(j)*b(j)*q,
- a(j)*d(j)*q + b(j)*c(j)*q + b(j)*c(j) + d(j)*a(j)*q,
q*(b(j)*d(j) + d(j)*b(j))),(0, - c(j)*d(j)*q + d(j)*c(j),
- c(j)*d(j)*q + d(j)*c(j),0))$
```

## 55. Gram-Schmidt Orthogonalisation Process

Let $V$ be a finite dimensional vector space over the real numbers with a positive definite scalar product $(,)$. We assume that $\dim V = n$. Suppose that $w_1, w_2, \ldots, w_n$ are a basis of $V$. We define

$$v_1 = w_1$$

$$v_2 = w_2 - \frac{(v_1, w_2)}{(v_1, v_1)} v_1$$

$$v_3 = w_3 - \frac{(v_2, w_3)}{(v_2, v_2)} v_2 - \frac{(v_1, w_3)}{(v_1, v_1)} v_1$$

$$\ldots = \ldots$$

$$v_n = w_n - \frac{(v_{n-1}, w_n)}{(v_{n-1}, v_{n-1})} v_{n-1} - \ldots - \frac{(v_1, w_n)}{(v_1, v_1)} v_1.$$

Then the unit vectors

$$u_j := \frac{v_j}{||v_j||},$$

where $||v_j|| := \sqrt{(v_j, v_j)}$, are mutually orthogonal and are an orthonormal basis of $V$. This is called the Gram-Schmidt orthogonalization process.

In the program we consider the Hilbert space $L_2[-1, 1]$. The scalar product is given by

$$(f, g) := \int\limits_{-1}^{1} f(x) g^*(x) dx.$$

A basis is given by

$$\{1, \ x, \ x^2, \ x^3, \ \ldots\}.$$

However, this basis is not orthogonal. We select the subset

$$\{w_0 = 1, \ w_1 = x, \ w_2 = x^2, \ w_3 = x^3, \ w_4 = x^4\}$$

and evaluate the corresponding $v_0, v_1, v_2, v_3, v_4$ to obtain the normalized Legendre polynomials.

*Remark:* The normalized Legendre polynomials form an orthonormal basis in the Hilbert space $L_2[-1, 1]$.

```
%schmidt.our;

n := 4;

% scalarproduct in L2(-1,1);
procedure sp(u,v);
sub(x=1,int(u*v,x))-sub(x=-1,int(u*v,x));

procedure schmidt(k);
begin
scalar s;
if k = 0 then return 1/2
else
s := x^k - for j := 0:k-1
sum sp(x^k,schmidt(j))*schmidt(j)/sp(schmidt(j),schmidt(j));
return s/sqrt(sp(s,s));
end;

for k := 0:n do
<<write " P(",k,") = ", schmidt(k)>>;
```

The output is

```
P(0) = 1/2$
P(1) = (SQRT(3)*X)/SQRT(2)$
P(2) = (SQRT(5)*(9/2*X**2 - 3/2))/(3*SQRT(2))$
P(3) = (SQRT(7)*X*(25/2*X**2 - 15/2))/(5*SQRT(2))$
P(4) = (3*(35*X**4 - 30*X**2 + 3))/(8*SQRT(2))$
```

## 56. Soliton Theory and Quantum Mechanics

Soliton theory and quantum mechanics are closely related. The starting point in soliton theory is the so-called *Lax pair* $L$ and $M$, where $L$ satisfies an eigenvalue equation and $M$ a time evolution equation. Here we give some applications of quantum mechanics in soliton theory.

Let $L$ and $M$ be two linear operators. Assume that

$$Lv = \lambda v \tag{1}$$

$$\frac{\partial v}{\partial t} = Mv, \tag{2}$$

where $L$ is the linear operator of the spectral problem (1), $M$ is the linear operator of an associated time evolution equation (2) and $\lambda$ is a real parameter. We find that

$$L_t v = [M, L]v, \tag{3}$$

where

$$[M, L] \equiv ML - LM \tag{4}$$

is the commutator and

$$L_t v = \frac{\partial (Lv)}{\partial t} - L\frac{\partial v}{\partial t}. \tag{5}$$

*Example:* We consider

$$L := \frac{\partial^2}{\partial x^2} + u(x, t) \tag{6}$$

and

$$M := 4\frac{\partial^3}{\partial x^3} + 6u(x, t)\frac{\partial}{\partial x} + 3\frac{\partial u}{\partial x}. \tag{7}$$

Equation (3) yields

$$\frac{\partial u}{\partial t} = 6u\frac{\partial u}{\partial x} + \frac{\partial^3 u}{\partial x^3}. \tag{8}$$

This is the Korteweg de Vries equation. In the program we evaluate the right hand side of (8).

```
%Lax.red;

operator L;
operator M;
operator v;
operator u;
depend v, x;
depend u, x;

L := df(v,x,2) + u*v;
M := 4*df(v,x,3) + 6*u*df(v,x) + 3*df(u,x)*v;

A := sub(v=L,M);
B := sub(v=M,L);
com := A - B;
```

The output is

```
A := 4*DF(U,X,3)*V + 12*DF(U,X,2)*DF(V,X) +
15*DF(U,X)*DF(V,X,2) + 9*DF(U,X)*U*V + 4*DF(V,X,5) +
10*DF(V,X,3)*U + 6*DF(V,X)*U**2$

B := 3*DF(U,X,3)*V + 12*DF(U,X,2)*DF(V,X) +
15*DF(U,X)*DF(V,X,2) + 3*DF(U,X)*U*V + 4*DF(V,X,5) +
10*DF(V,X,3)*U + 6*DF(V,X)*U**2$

COM := V*(DF(U,X,3) + 6*DF(U,X)*U)$
```

In our second example, we consider the linear operator (pseudo differential operator)

$$L := \partial + u_2(\mathbf{x})\partial^{-1} + u_3(\mathbf{x})\partial^{-2} + u_4(\mathbf{x})\partial^{-3} + \ldots, \tag{1}$$

where $\partial = \partial/\partial x_1$ and

$$\partial^{-1}f(\mathbf{x}) = \int^{x_1} f(s, x_2, \ldots, x_n)ds. \tag{2}$$

Let $B_n$ be the differential part of the operator $(L(\mathbf{x}, \partial))^n$. The program determines $B_n$. We obtain

$$B_1 = \partial \tag{3}$$

$$B_2 = \partial^2 + 2u_2 \tag{4}$$

$$B_3 = \partial^3 + 3u_2\partial + 3u_3 + 3\frac{\partial u_2}{\partial x_1} \tag{5}$$

$$B_4 = \partial^4 + 4u_2\partial^2 + \left(4u_3 + 6\frac{\partial u_2}{\partial x_1}\right)\partial + 4u_4 + 6\frac{\partial u_3}{\partial x_1} + 4\frac{\partial^2 u_2}{\partial x_1^2} + 6u_2^2. \tag{6}$$

In the program we evaluate $B_2$, $B_3$ and $B_4$. We have to implement the rules

$$\partial(\partial^{-1})\psi = \psi, \qquad \partial^{-1}(\partial\psi) = \psi, \tag{7}$$

where the quantity $\psi$ is denoted by $p$ in the program.

```
%pseudo.red;

operator R;
depend u2, x1;
depend u3, x1;
depend u4, x1;
depend p, x1;

let df(R(p),x1) = p;
let df(R(R(p)),x1,2) = p;
let df(R(R(R(p))),x1,3) = p;
let df(R(R(R(R(p)))),x1,4) = p;
let R(df(p,x1)) = p;
let R(R(df(p,x1,2))) = p;
let R(R(R(df(p,x1,3)))) = p;
let R(R(R(R(df(p,x1,4))))) = p;
for all A let R(df(p,x1)*A) = p*A;
for all A, B let R(A+B) = R(A) + R(B);

L := df(p,x1) + u2*R(p) + u3*R(R(p)) + u4*R(R(R(p))));
L2 := sub(p=L,L)$
L3 := sub(p=L,L2)$
L4 := sub(p=L,L3)$
for all x let R(x) = 0;
B2 := L2;
B3 := L3;
B4 := L4;
```

The output is

```
B2 := DF(P,X1,2) + 2*P*U2$
B3 := DF(P,X1,3) + 3*DF(P,X1)*U2 + 3*DF(U2,X1)*P + 3*P*U3$
B4 := DF(P,X1,4) + 4*DF(P,X1,2)*U2 + 6*DF(P,X1)*DF(U2,X1) +
4*DF(P,X1)*U3 + 4*DF(U2,X1,2)*P + 6*DF(U3,X1)*P + 6*P*U2**2 +
4*P*U4$
```

In our third example we consider hierarchies of soliton equations. Hierarchies of soliton equations can be derived with the help of pseudo differential operators. Consider

$$L(\mathbf{x}, \partial) := \partial + u_2(\mathbf{x})\partial^{-1} + u_3(\mathbf{x})\partial^{-2} + u_4(\mathbf{x})\partial^{-3} + \ldots, \tag{1}$$

where $u_n$ are functions in infinitely many "time" variables $\mathbf{x} = (x_1, x_2, \ldots)$ and $\partial$ denotes $\partial/\partial x_1$. Let $B_n(\mathbf{x}, \partial)$ $(n = 1, 2, \ldots)$ be the differential part of $(L(\mathbf{x}, \partial))^n$. The first four of the $B_n's$ are given by

$$B_1 = \partial \tag{2a}$$

$$B_2 = \partial^2 + 2u_2 \tag{2b}$$

$$B_3 = \partial^3 + 3u_2\partial + 3u_3 + 3\frac{\partial u_2}{\partial x_1} \tag{2c}$$

$$B_4 = \partial^4 + 4u_2\partial^2 + \left(4u_3 + 6\frac{\partial u_2}{\partial x_1}\right)\partial + 4u_4 + 6\frac{\partial u_3}{\partial x_1} + 4\frac{\partial^2 u_2}{\partial x_1^2} + 6u_2^2. \tag{2d}$$

Consider a system of linear equations for an eigenfunction $w$, namely

$$L(\mathbf{x}, \partial)w(\mathbf{x}, k) = kw(\mathbf{x}, k) \tag{3}$$

$$\frac{\partial w(\mathbf{x}, k)}{\partial x_n} = B_n(\mathbf{x}, \partial)w(\mathbf{x}, k), \tag{4}$$

where $n = 1, 2, \ldots$. From the compatibility condition of (3) and (4) we find

$$\frac{\partial L}{\partial x_n} = [B_n, L] = B_n L - L B_n \tag{5}$$

or equivalently

$$\frac{\partial B_m}{\partial x_n} - \frac{\partial B_n}{\partial x_m} = [B_m, B_n]. \tag{6}$$

We evaluate (6) for $B_3$ and $B_4$.

```
%soliton.red;
on1$

operator B3, B4;
operator A, B, W, Z;
operator u2, u3, u4, p;
depend u2, x1, x2, x3, x4;
depend u3, x1, x2, x3, x4;
depend u4, x1, x2, x3, x4;
depend p, x1;
depend A, x1, x2, x3, x4;
depend B, x1, x2, x3, x4;
depend W, x1, x2, x3, x4;
depend Z, x1, x2, x3, x4;
B4 := df(p,x1,4) + 4*u2*df(p,x1,2) + (4*u3 + 6*df(u2,x1))*df(p,x1)
+ 4*u4*p + 6*df(u3,x1)*p + 4*df(u2,x1,2)*p + 6*u2*u2*p;
B3 := df(p,x1,3) + 3*u2*df(p,x1) + 3*u3*p + 3*df(u2,x1)*p;
A := df(B3,x4);
B := df(B4,x3);
W := sub(p=B3,B4);
Z := sub(p=B4,B3);
R := B-A+W-Z;
c0 := coeffn(R,df(p,x1,7),1);
c1 := coeffn(R,df(p,x1,6),1);
c2 := coeffn(R,df(p,x1,5),1);
c3 := coeffn(R,df(p,x1,4),1);
c4 := coeffn(R,df(p,x1,3),1);
c5 := coeffn(R,df(p,x1,2),1);
c6 := coeffn(R,df(p,x1),1);
c7 := coeffn(R,p,1);
```

The output is

```
C0 := 0$ C1 := 0$ C2 := 0$ C3 := 0$ C4 := 0$
C5 := 4*( - DF(U2,X1,3) - 6*DF(U2,X1)*U2 + DF(U2,X3) -
3*DF(U3,X1,2) - 3*DF(U4,X1))$
C6 := 6*DF(U2,X1,X3) - 3*DF(U2,X1,4) - 18*DF(U2,X1,2)*U2 -
18*DF(U2,X1)**2 + 12*DF(U2,X1)*U3 - 3*DF(U2,X4) - 10*DF(U3,X1,3) +
12*DF(U3,X1)*U2 + 4*DF(U3,X3) - 12*DF(U4,X1,2)$
C7 :=  - 3*DF(U2,X1,X4) - DF(U2,X1,5) - 12*DF(U2,X1,3)*U2 +
4*DF(U2,X1,2,X3) - 18*DF(U2,X1,2)*DF(U2,X1) + 12*DF(U2,X1,2)*U3 +
18*DF(U2,X1)*DF(U3,X1) - 36*DF(U2,X1)*U2**2 + 12*DF(U2,X3)*U2 +
6*DF(U3,X1,X3) - 3*DF(U3,X1,4) - 6*DF(U3,X1,2)*U2 + 12*DF(U3,X1)*U3 -
3*DF(U3,X4) - 4*DF(U4,X1,3) - 12*DF(U4,X1)*U2 + 4*DF(U4,X3)$
```

## 57. Padé Approximation

Let $f$ be an analytic function. The function $f$ can be expanded into a power series

$$f(x) = a_0 + a_1 x + a_2 x^2 + \ldots \qquad . \tag{1}$$

We define

$$[L, L] := \frac{\det \begin{vmatrix} a_1 & a_2 & \cdots & a_{L+1} \\ \vdots & \vdots & & \\ a_L & a_{L+1} & \cdots & a_{L+L} \\ \sum_{j=L}^{L} a_{j-L} x^j & \sum_{j=L-1}^{L} a_{j-L+1} x^j & \cdots & \sum_{j=0}^{L} a_j x^j \end{vmatrix}}{\det \begin{vmatrix} a_1 & a_2 & \cdots & a_{L+1} \\ \vdots & \vdots & & \\ a_L & a_{L+1} & \cdots & a_{L+L} \\ x^L & x^{L-1} & \cdots & 1 \end{vmatrix}} \tag{2}$$

where $L$ is a positive integer. Then $\hat{f}_L := [L, L]$ is called the Padé approximation of $f$.

In the program we calculate the Padé approximation $[2, 2]$ for $f(x) = \sin(x)$, where

$$[2, 2] = \frac{\det \begin{vmatrix} a_1 & a_2 & a_3 \\ a_2 & a_3 & a_4 \\ a_0 x^2 & a_0 x + a_1 x^2 & a_0 + a_1 x + a_2 x^2 \end{vmatrix}}{\det \begin{vmatrix} a_1 & a_2 & a_3 \\ a_2 & a_3 & a_4 \\ x^2 & x & 1 \end{vmatrix}} \tag{3}$$

where $a_0 = 0$, $a_1 = 1$, $a_2 = 0$, $a_3 = -1/6$ and $a_4 = 0$. We find that

$$[2, 2] = \frac{x}{1 + \dfrac{x^2}{6}} . \tag{4}$$

An application in quantum mechanics is as follows: consider the Hamilton operator

$$\hat{H} = -\frac{\hbar^2}{2m} \frac{1}{r^2} \frac{d}{dr} \left( r^2 \frac{d}{dr} \right) + \lambda V(r). \tag{5}$$

A scalar product of two real functions $\alpha(r)$, $\beta(r)$ is to be defined by

$$\langle \alpha | \beta \rangle := \int_0^\infty \alpha(r)\beta(r)r^2 dr. \tag{6}$$

To describe scattering at a given energy $E = k^2\hbar^2/2m$, we need a wave function $u(r)$, bounded at $r = 0$, which satisfies $(E - \hat{H})|u\rangle = 0$ and has the asymptotic from

$$u(r) \sim \frac{1}{kr}(\sin kr + \tan\delta\cos kr), \quad \text{for } r \to \infty \tag{7}$$

where $\delta$ is the phase shift. The function $|u\rangle$ may also be described as the solution of the *Lippmann-Schwinger integral equation*

$$|u\rangle = |0\rangle + \lambda G_0 V|u\rangle, \tag{8}$$

where the Green's function $G_0$ may be written as

$$\langle r|G_0|r'\rangle = -2mkj_0(kr_<)n_0(kr_>) \tag{9}$$

and the state $|0\rangle$ represents the function $(\sin kr)/kr$. An expansion for $|u\rangle$ in powers of the potential strength $\lambda$ is obtained by iterating (8)

$$|u\rangle = \sum_{i=0}^\infty \lambda^i |i\rangle \tag{10}$$

where the state $|i\rangle$ is defined as

$$|i\rangle = (G_0 V)^i |0\rangle. \tag{11}$$

Inserting (10) into the relation between $\tan\delta$ and $|u\rangle$, which is

$$t \equiv -\frac{\tan\delta}{2mk\lambda} = \langle 0|V|u\rangle, \tag{12}$$

we find the Born series

$$t = \sum_{i=0}^\infty a_i \lambda^i, \tag{13}$$

where $a_i = \langle 0|V|i\rangle$. For large enough values of $|\lambda|$, these series will diverge. A method of obtaining numerical solutions to problems in scattering theory is that of stationary variational principles. For a small variation of $|\phi\rangle$ and $|\phi'\rangle$ about $|u\rangle$, the quantity $I$ is stationary, and at $|\phi\rangle = |\phi'\rangle = |u\rangle$ takes on the value shown in brackets:

$$I = [t] = \langle 0|V|\phi\rangle + \langle \phi'|V|0\rangle - \langle \phi'|(V - \lambda V G_0 V)|\phi\rangle. \tag{14}$$

We take $|\phi\rangle = |\phi'\rangle$ in (14). In practice, we take an approximation to $|u\rangle$, namely $|\phi\rangle = |u(N)\rangle$, which depends on $N$ parameters and calculate the expression $I$. The parameters are then varied until $I$ is stationary, thus determining the approximate $|u(N)\rangle$. It is simplest to assume that the dependence on the parameters is linear, and a natural choice for $|u(N)\rangle$, bearing (14) in mind, is

$$|u(N)\rangle = \sum_{i=0}^{N-1} c_i |i\rangle, \tag{15}$$

where the real coefficients $c_i$ are to be varied. We find that $I$ is given by

$$I = -\left(\sum_{i,j=0}^{N-1} c_i c_j (a_{i+j} - \lambda a_{i+j+1}) - 2 \sum_{i=0}^{N-1} c_i a_i\right) = -\left(c^T M c - 2c^T a\right) \tag{16}$$

in a matrix notation with

$$M_{ij} = a_{i+j} - \lambda a_{i+j+1}. \tag{17}$$

The form (16) is stationary when $c$ is given by

$$c = M^{-1}a, \tag{18}$$

and here $I$, which leads to an estimate for $\tan \delta$ accurate to the second order, is

$$I^{(N)} = t^{(N)} = a^T M^{-1} a. \tag{19}$$

We find that $I^{(N)}$ is just the $[N, N-1]$ Padé approximant to $t$.

```
%Pade.our;

%Procedure for the Taylor expansion;
procedure tay(u,x,n);
  begin scalar ser,fac;
  ser := sub(x=0,u); fac := 1;
  for j:=1:n do
  <<u:=df(u,x); fac:=fac*j;
  ser:=ser+sub(x=0,u)*x**j/fac>>;
  return(ser);
end;

% We evaluate the Pade approximation for sin(x);
operator a;
P := tay(sin(x),x,5);
for j:=0:4 do
begin
a(j) := coeffn(P,x,j);
end;
a(0); a(1); a(2); a(3); a(4);

size := 3;  % size of the matrices;
matrix M(size,size);
for j:=1:size do
begin
M(1,j):=a(j);
end;
M(2,1)  := a(2); M(2,2):=a(3); M(2,3):=a(4);
M(3,1)  := a(0)*x**2; M(3,2):=a(0)*x+a(1)*x**2;
M(3,3)  := for j:=0 step 1 until 2 sum a(j)*x**j;

matrix N(3,3);
for j:=1:size do
begin
N(1,j):=a(j);
end;
N(2,1):=a(2); N(2,2):=a(3); N(2,3):=a(4);
N(3,1):=x**2; N(3,2):=x; N(3,3):= 1;

write "[2,2] =";  det(M)/det(N);
```

The output is

```
(6*x)/(x**2 + 6)$
```

## 58. Cumulant Expansion

The cumulant expansion plays a central role in statistical physics. Let $\lambda$ be a real parameter and $a_n, b_n \in \mathcal{R}$. Assume that

$$\exp\left[\sum_{n=1}^{\infty} \frac{\lambda^n b_n}{n!}\right] = \sum_{n=0}^{\infty} \frac{\lambda^n a_n}{n!}, \tag{1}$$

where $a_0 = 1$. We determine the relation between $a_n$ and $b_n$. An arbitrary term of the exponential function of the left hand side of (1) is given by

$$\frac{1}{k}\left(\sum_{n=1}^{\infty} \frac{\lambda^n b_n}{n!}\right) = \frac{1}{k!}\left(\sum_{n_1=1}^{\infty} \frac{\lambda^{n_1} b_{n_1}}{n_1!}\right) \cdots \left(\sum_{n_k=1}^{\infty} \frac{\lambda^{n_k}}{n_k!}\right)$$

$$= \frac{1}{k!} \sum_{n_1=1}^{\infty} \sum_{n_2=1}^{\infty} \cdots \sum_{n_k=1}^{\infty} \frac{\lambda^{n_1+n_2+\cdots+n_k} b_{n_1} b_{n_2} \cdots b_{n_k}}{n_1! n_2! \cdots n_k!}. \tag{2}$$

Therefore

$$\exp\left[\sum_{n=1}^{\infty} \frac{\lambda^n b_n}{n!}\right] = 1 + \sum_{n=1}^{\infty} \frac{\lambda^n b_n}{n!} + \frac{1}{2!} \sum_{n_1=1}^{\infty} \sum_{n_2=1}^{\infty} \frac{\lambda^{n_1+n_2} b_{n_1} b_{n_2}}{n_1! n_2!} + \cdots$$

$$= \frac{1}{k!} \sum_{n_1=1}^{\infty} \sum_{n_2=1}^{\infty} \cdots \sum_{n_k=1}^{\infty} \frac{\lambda^{n_1+n_2+\cdots+n_k} b_{n_1} b_{n_2} \cdots b_{n_k}}{n_1! n_2! \cdots n_k!} = \sum_{n=0}^{\infty} \frac{\lambda^n a_n}{n!}. \tag{3}$$

Equating terms of the same order in $\lambda$, we obtain for the first three terms

$$\lambda^1 : a_1 = b_1, \quad \lambda^2 : a_2 = b_2 + b_1^2, \quad \lambda^3 : a_3 = b_3 + 3b_2 b_1 + b_1^3. \tag{4}$$

It follows that

$$b_1 = a_1, \quad b_2 = a_2 - a_1^2, \quad b_3 = a_3 - 3a_2 a_1 + 2a_1^3. \tag{5}$$

```
%cumul.red;

depend P, x;
operator a;
for n:=1:8 sum (x**n)*a(n)/(for j:=1:n product j);
P := a(0) + ws;

procedure tay(u,x,n);
begin
  scalar ser,fac;
  ser := sub(x=0,u); fac := 1;
  for j:=1:n do
  <<u:=df(u,x); fac:=fac*j;
  ser:=ser+sub(x=0,u)*x**j/fac>>;
  return(ser);
end;

operator b;
for n:=1:8 sum (x**n)*b(n)/(for j:=1:n product j);
expan := tay(exp(ws),x,5);

%Example 1;
coeffn(expan,x,1);
coeffn(P,x,1);

%Example 2;
coeffn(expan,x,2);
coeffn(P,x,2);

%Example 3;
coeffn(expan,x,3);
coeffn(P,x,3);
```

The output is

```
b(1)$
a(1)$

(b(2) + b(1)**2)/2$
a(2)/2$

(b(3) + 3*b(2)*b(1) + b(1)**3)/6$
a(3)/6$
```

## 59. Leverrier's Method

The characteristic polynomial of an $n \times n$ matrix $A$ will be a polynomial of degree $n$ in $\lambda$. Hence, it may be written as

$$P(\lambda) := \det(\lambda I - A) = \lambda^n - c_1 \lambda^{n-1} - c_2 \lambda^{n-2} - \ldots - c_{n-1} \lambda - c_n. \tag{1}$$

We now present Leverrier's method to find the coefficients, $c_i$, of the polynomial. It is fairly insensitive to the individual peculiarities of the matrix $A$. The method has an added advantage that the inverse of $A$, if it exists, is also obtained in the process of determining the coefficients, $c_i$. Obviously we also obtain the determinant. One may then utilize both $A$ and $A^{-1}$ for checking the accuracy of individual eigenvalues using the power or the inverse power method.

*Definition:* The *trace* of an $n \times n$ matrix $A$, denoted as $\text{tr}(A)$, is the sum of the elements on the main diagonal of $A$,

$$\text{tr}(A) := \sum_{i=1}^{n} a_{ii}. \tag{2}$$

The trace is related to the eigenvalues of $A$ by

$$\text{tr}(A) = \sum_{i=1}^{n} \lambda_i, \tag{3}$$

where $\lambda_1, \lambda_2, \ldots, \lambda_n$ are the eigenvalues.

The Leverrier method determines the coefficients, $c_i$, of $P(\lambda)$ by obtaining the trace of each of the matrices, $B_1, B_2, \ldots, B_n$, generated as follows. Set $B_1 = A$ and compute $c_1 = \text{tr}(B_1)$. Then we compute

$$B_k = A(B_{k-1} - c_{k-1}I), \qquad c_k = \left(\frac{1}{k}\right)\text{tr}(B_k), \qquad k = 2, 3, \ldots, n. \tag{4}$$

The inverse of a nonsingular matrix $A$ can be obtained from the relationship

$$A^{-1} = \left(\frac{1}{c_n}\right)(B_{n-1} - c_{n-1}I). \tag{5}$$

```
%lever.red;

n := 4;
matrix A(n,n);
A(1,1) := 5; A(1,2) := 4; A(1,3) := 1; A(1,4) := 1;
A(2,1) := 4; A(2,2) := 5; A(2,3) := 1; A(2,4) := 1;
A(3,1) := 1; A(3,2) := 1; A(3,3) := 4; A(3,4) := 2;
A(4,1) := 1; A(4,2) :=1 ; A(4,3) := 2; A(4,4) := 4;

matrix ID(n,n);
ID(1,1) := 1; ID(1,2) := 0; ID(1,3) := 0; ID(1,4) := 0;
ID(2,1) := 0; ID(2,2) := 1; ID(2,3) := 0; ID(2,4) := 0;
ID(3,1) := 0; ID(3,2) := 0; ID(3,3) := 1; ID(3,4) := 0;
ID(4,1) := 0; ID(4,2) := 0; ID(4,3) := 0; ID(4,4) := 1;

matrix c(n,1); matrix B(n,n); matrix BI(n,n); matrix D(n,n);
B := A; c(1,1) := trace(A);

for k:=2:n do
begin
D := A*(B - c(k-1,1)*ID);
c(k,1) := (1/k)*trace(D);
if k=(n-1) then BI := D;
B := D;
end;

% coefficients of the characteristic polynomial;
for j:=1:n do write c(j,1);

%determinant;
write "determinant = " , -c(n,1);

%inverse matrix;
matrix AI(n,n);
if c(n,1) = 0 then write "inverse does not exist" else
if c(n,1) neq 0 then
AI := (1/c(n,1))*(BI - c(n-1,1)*ID);
```

The output is

```
c(1,1) := 18$  18$ -97$ 180$ -100$  determinant = 100$
mat((14/25,( - 11)/25,( - 1)/50,( - 1)/50),(( - 11)/25,14/25,
(-1)/50,( - 1)/50),(( - 1)/50,( - 1)/50,17/50,( - 4)/25),(( - 1)/
50,( - 1)/50,( - 4)/25,17/50))$
```

## MATHEMATICA PROGRAMS

Shown below is a collection of MATHEMATICA programs for our quantum mechanical problems.

First we consider problem 2. We show that

$$\psi(x,t) = \frac{B}{(1 + i\hbar t/ma^2)^{1/2}} \exp\left(-\frac{x^2}{2a^2(1 + i\hbar t/ma^2)}\right)$$

satisfies the Schrödinger equation (1) (see problem 2 and equation (2) with $k_0 = 0$).

```
(* wavepack.ma *)
f1 = B/Sqrt[1 + I*hb*t/(m*a2)]
f2 = Exp[-x*x/(2*a2*(1+I*hb*t/(m*a2)))]
psi = f1*f2
res = I*hb*D[psi,t] + hb*hb/(2*m)*D[psi,x,x]
res = Simplify[res]
```

In our second program we consider problem 5. The trial function $u$ is inserted into (4) and the expectation value of the energy is minimalized with respect the the parameter $a$.

```
(* trial.ma *)
u = x*Exp[-a*x]
res1 = -hb*hb/(2*m)*D[u,x,x] + c*x*u
res2 = Integrate[u*res1,x]
res3 = -res2 /. x -> 0
res4 = Integrate[u*u,x]
norm = res4 /. x -> 0
expe = res3/norm
minim = D[expe,a]
res5 = Solve[minim==0,a]
```

The output is

```
{{a -> ((3/2)^(1/3)*c^(1/3)*m^(1/3))/hb^(2/3)},
 {a -> ((-1)^(2/3)*(3/2)^(1/3)*c^(1/3)*m^(1/3))/hb^(2/3)},
 {a -> ((-1)^(4/3)*(3/2)^(1/3)*c^(1/3)*m^(1/3))/hb^(2/3)}}
```

We consider the spin-1 matrices (see problem 16), where we evaluate the commutator of the two matrices $s_+$ and $s_-$ and then determine the eigenvalues of the commutator.

```
sp = {{0,Sqrt[2]*hb,0},{0,0,Sqrt[2]*hb},{0,0,0}}
sm = {{0,0,0},{Sqrt[2]*hb,0,0},{0,Sqrt[2]*hb,0}}
comm = sp . sm - sm . sp
Eigenvalues[comm]
```

The output is

```
      2         2
{2 hb , -2 hb , 0}
```

Finally we consider the gamma matrices (see problem 49) and the Kronecker product (see problem 26).

```
sx = {{0,1},{1,0}}
sy = {{0,-I},{I,0}}
sz = {{1,0},{0,-1}}
g1 = Outer[Times, sy, sx]
g2 = Outer[Times, sy, sy]
g3 = Outer[Times, sy, sz]
g4 = Outer[Times, sz, IdentityMatrix[2]]
```

The output is

```
{{{{0, 0}, {0, 0}}, {{0, -I}, {-I, 0}}},
 {{{0, I}, {I, 0}}, {{0, 0}, {0, 0}}}}

{{{{0, 0}, {0, 0}}, {{0, -1}, {1, 0}}},
 {{{0, 1}, {-1, 0}}, {{0, 0}, {0, 0}}}}

{{{{0, 0}, {0, 0}}, {{-I, 0}, {0, I}}},
 {{{I, 0}, {0, -I}}, {{0, 0}, {0, 0}}}}

{{{{1, 0}, {0, 1}}, {{0, 0}, {0, 0}}},
 {{{0, 0}, {0, 0}}, {{-1, 0}, {0, -1}}}}
```

## MAPLE PROGRAMS

Here follows a collection of MAPLE programs for our quantum mechanical problems.

First we consider problem 2. We show that

$$\psi(x,t) = \frac{B}{(1 + i\hbar t/ma^2)^{1/2}} \exp\left(-\frac{x^2}{2a^2(1 + i\hbar t/ma^2)}\right)$$

satisfies the Schrödinger equation (1) (see problem 2 and equation (2) with $k_0 = 0$).

```
# wavpack.map;
f1 := B/(sqrt(1+I*hb*t/(m*a2)));
f2 := exp(-x*x/(2*a2*(1+I*hb*t/(m*a2))));
psi := f1*f2;
res1 := hb*hb*diff(psi,x,x)/(2*m);
res2 := I*hb*diff(psi,t);
result1 := res1 + res2;
result2 := result1/f2;
result3 := simplify(result2);
```

In our second program we consider problem 5. The trial function $u$ is inserted into (4) and the expectation value of the energy is minimalized with respect the the parameter $a$.

```
# trial.map;
u := x*exp(-a*x);
res1 := -hb*hb/(2*m)*diff(u,x,x) + c*x*u;
res2 := int(u*res1,x);
res3 := -subs(x=0,res2);
res4 := int(u*u,x);
norm := -subs(x=0,res4);

expe := res3/norm;
minim := diff(expe,a);
res5 := solve(minim=0,a);
```

Next we consider the spin-1 matrices (see problem 16), where we evaluate the commutator of the matrices $s_+$ and $s_-$ and then determine the eigenvalues of the commutator. We give two different implementations.

```
# spin1.map;
with(linalg):
sp:=matrix(3,3,[0,sqrt(2)*hb,0,0,0,sqrt(2)*hb,0,0,0]);
sm:=matrix(3,3,[0,0,0,sqrt(2)*hb,0,0,0,sqrt(2)*hb,0]);
comm := evalm(sp&*sm - sm&*sp);

# an alternative option is;
sp := array(1..3,1..3);
sp[1,1] := 0; sp[1,2] := sqrt(2)*hb; sp[1,3] := 0;
sp[2,1] := 0; sp[2,2] := 0; sp[2,3] := sqrt(2)*hb;
sp[3,1] := 0; sp[3,2] := 0; sp[3,3] := 0;
sm := array(1..3,1..3);
sm[1,1] := 0; sm[1,2] := 0; sm[1,3] := 0;
sm[2,1] := sqrt(2)*hb; sm[2,2] := 0; sm[2,3] := 0;
sm[3,1] := 0; sm[3,2] := sqrt(2)*hb; sm[3,3] := 0;
comm := evalm(sp&*sm - sm&*sp);
result := eigenvals(comm);
```

The Hermite, Laguerre, Legendre, and Chebyshev polynomials are built-in functions in MAPLE. The following MAPLE program sums the first four Legendre polynomials. It then shows that $P(4,x)$ satisfies the Legendre differential equation with $n = 4$ (see problem 41).

```
#Legend.map
with(orthopoly):
r := 0;
for j from 0 by 1 to 3 do
r := r + P(j,x)
od;

n := 4;
result := (1-x*x)*diff(P(n,x),x,x) - 2*x*diff(P(n,x),x) +
n*(n+1)*P(n,x);

simplify(result);
```

Finally, we give a procedure for the Chebyshev polynomials.

```
#procedure for Chebyshev polynomials, cheby.map
Chebyshev := proc(n)
local p, k;
p[0] := 1; p[1] := x;
if n<=1 then
RETURN(eval(p)) fi;
for k from 2 to n do
p[k] := expand(2*x*p[k-1]-p[k-2]);
od:
RETURN(eval(p))
end:
```

The Chebyshev polynomials up to degree 4 are now called by $Chebyshev(4)$.

## C++ PROGRAMS

Here we give three C++ programs. In the first program we solve the equation of motion of the driven two level system (see problem 24). In the second program we find the Chebyshev polynomials up to degree 6. The third program implements Bose operators.

In the first program we apply a Runge-Kutta technique for explicitly time-dependent systems of first order ordinary differential equations.

```cpp
// KUTTA4.CPP

#include <iostream.h>    // for cout
#include <math.h>        // for cos(), sin()

#define      N           4       // System Order

typedef  double  vector[N];
typedef  double  scalar;

void F(scalar h, scalar t, vector x, vector hf);

void main()
{
  unsigned T = 2000;  // number of steps, total time = T*h
  unsigned  i, j, k, l;
  scalar    h = 0.01, t = 0.0, tk;
  vector    xk, x = { 1.0, 0.0, 0.0, 0.0 }; // Initial conditions
  scalar    a[6] = { 0.0, 1.0/4.0, 3.0/8.0, 12.0/13.0, 1.0, 1.0/2.0 };
  scalar    c[6] = { 16.0/135.0, 0.0, 6656.0/12825.0, 28561.0/56430.0,
    -9.0/50.0, 2.0/55.0 };
  scalar    b[6][5] =
    { {0.0},  {1.0/4.0},  {3.0/32.0, 9.0/32.0},
      {1932.0/2197.0, -7200.0/2197.0, 7296.0/2197.0},
      {439.0/216.0, -8.0, 3680.0/513.0, -845.0/4104.0},
      {-8.0/27.0, 2.0, -3544.0/2565.0, 1859.0/4104.0, -11.0/40.0}
    };
  vector f[6];
```

```
   for (i=0; i<T; i++)
   {
     t += h;
     F(h, t, x, f[0]);

     for (k=1; k<=5; k++)
     {
       tk = t + a[k]*h;
       for (l=0; l<N; l++)
       {
xk[l] = x[l];
for (j=0; j<=k-1; j++) xk[l] += b[k][j]*f[j][l];
       }
       F(h, tk, xk, f[k]);
     }

     for (l=0; l<N; l++)
     for (k=0; k<6; k++) x[l] += c[k]*f[k][l];

   }
   cout << " total time = " << T*h << "\n";
   cout << " prob1 = " << x[0]*x[0] + x[1]*x[1] << "\n";
   cout << " prob2 = " << x[2]*x[2] + x[3]*x[3] << "\n";
}

void F(scalar h, scalar t, vector x, vector hf)
{
   scalar w11 = 0.5; scalar w12 = 0.5;
   scalar w21 = 0.5; scalar w22 = 0.5;
   scalar om = 0.1;
   hf[0] = h*(cos(t)*w11*x[1]+cos(t)*cos(om*t)*w12*x[3]
   -cos(t)*sin(om*t)*w12*x[2]);
   hf[1] = h*(-cos(t)*w11*x[0]-cos(t)*cos(om*t)*w12*x[2]
   -cos(t)*sin(om*t)*w12*x[3]);
   hf[2] = h*(cos(t)*cos(om*t)*w21*x[1]+cos(t)*sin(om*t)*w21*x[0]
   +cos(t)*w22*x[3]);
   hf[3] = h*(-cos(t)*cos(om*t)*w21*x[0]+cos(t)*sin(om*t)*w21*x[1]
   -cos(t)*w22*x[2]);
}
```

In our second program we evaluate the Chebyshev polynomials up to degree 6 (see problem 44). Then we calculate the numerical value of the Chebyshev polynomial of degree 6 for $x = 2.0$.

```cpp
// cheby1.cpp

#include <iostream.h>      // for cout
#include <math.h>          // for pow()
#include <conio.h>         // for clrscr()

// evaluation of Chebyshev polynomials up to degree 6
// degree = N - 1

#define N 7

int tm[N]; int tm1[N]; int tm2[N];

void ccheby(int tm1[],int tm2[],int tm[])
{
    for (int i=(N-1);i>=1;i--)
    tm[i] = 2*tm1[i-1]-tm2[i];
    tm[0] = -tm2[0];
}

void print(int tm[])
{
 //    cout << "\n";
    for(int i=(N-1);i>=0;i--)
    if (tm[i] != 0)
    {
      if (tm[i] > 0) cout << "+";
    cout << tm[i] << "*x^" << i;
    }
    cout << "\n";
}
```

```
void main()
{
  clrscr();
  int i;
  for (i=0;i<N;i++) { tm1[i] = 0; }
  for (i=0;i<N;i++) { tm2[i] = tm1[i]; }
  tm2[0] = 1; tm1[1] = 1;
  print(tm2);
  print(tm1);
  for(i=2;i<N;i++)
  {
  ccheby(tm1,tm2,tm);
  print(tm);
  int j;
  for(j=0;j<N;j++)  { tm2[j] = tm1[j]; }
  for(j=0;j<N;j++)  { tm1[j] = tm[j];  }
  }

  // Evaluation of Cheby6(x) for x = 2.0
  double x = 2.0;
  double sum = 0.0;
  for (i=0;i<N;i++)
  {
  sum += tm[i]*pow(x,i);
  }
  cout << "\n" << "t(" << N-1 << ") = " << sum;
}
```

Finally we give a C++ program for Bose operators. Given the state $b^\dagger b^\dagger |0\rangle$ and the operator $b^\dagger bb$ the new state $2b^\dagger |0\rangle$ is calculated.

```cpp
// Bose.cpp

#include <iostream.h>    // for cout
#include <conio.h>       // for clrscr()

  // Definition of class State
class State
{
private:
   int m;
   int factor;
public:
   State();
   void bose_state(int cd);  // member function
   void display();           // member function
};

  // Implementation of class State
State::State()
{
   m = 0; factor = 1;
}

void State::bose_state(int cd)
{
   if (cd == -1) factor *= m--; else m++;
}

void State::display()
{
   if (factor == 0) cout << "0";
   else { cout << "+(" << factor << ")*";
   if (m!=0) { cout << "(";
   for (int i=0;i<m;i++) cout << "b+";
 cout << ")";
    }
 cout << "|0>";
 }
}       // end implementation class State
```

```
class Bose
{
public:
    State operator+(State& st2)
    {
    State st1 = st2;
    st1.bose_state(1);
    return st1;
    }
    State operator-(State& st2)
    {
    State st1 = st2;
    st1.bose_state(-1);
    return st1;
    }
};

void main()
{
    clrscr();
    State g;  // g is object of class State
    Bose b;   // b is object of class Bose
    g =  b+ (b- (b- (b+ (b+ g))));
    cout << "g = ";
    g.display();  // The dot . is the class object selector
}
```

# REFERENCES

## Books about Quantum Mechanics

Constantinescu, F. and Magyari, E., *Problems in Quantum Mechanics*, Pergamon, 1976

Cronin, J. A., Greenberg, D. F. and Telegdi, V. L., *University of Chicago Graduate Problems in Physics*, Addison-Wesley, Reading 1967

Flügge, S., *Practical Quantum Mechanics*, Springer-Verlag, 1974

Gasiorowicz, S., *Quantum Physics*, John Wiley, New York, 1974

Messiah, A., *Quantum Mechanics*, Interscience, New York, 1961

Prugovečki, E., *Quantum Mechanics in Hilbert Space*, Second Edition, Academic Press, New York, 1972

Steeb W.-H., *A Handbook of Terms used in Chaos and Quantumchaos*, Bibliographisches Institut, Mannheim, 1991

Steeb W.-H. and Louw J., *Chaos and Quantum Chaos*, World Scientific, Singapore, 1986

Steeb W.-H., *Kronecker product of matrices and applications*, Bibliographisches Institut, Mannheim, 1991

Steeb W.-H., *Hilbert Spaces, Generalized Functions and Quantum Mechanics*, Bibliographisches Institut, Mannheim, 1992

Steeb W.-H. *Chaos und Quanten Chaos in dynamischen Systemen*, Bibliographisches Institut, Mannheim, 1994

## Books about REDUCE, MATHEMATICA, MAPLE, C and C++

Char, B. M., Geddes, K. O., Gonnet, G. H., Leong, B. L., Monagan, M. B., and Watt, S. M., *Maple V Library Reference Manual*, Springer-Verlag, New York 1991

Char, B. M., Geddes, K. O., Gonnet, G. H., Leong, B. L., Monagan, M. B., and Watt, S. M., *Maple V Language Reference Manual*, Springer-Verlag, New York 1991

Crandall, R. E., *Mathematica for the Sciences*, Addison-Wesley, Redwood City Hearn A., *REDUCE User's Manual, Version 3.4*, The RAND Corporation, Santa Monica (CA) 1991

Hehl, F. W., Winkelmann, V. and Meyer, H., *Computer-Algebra*, Springer-Verlag, Berlin 1992

Holmes, M. H., Ecker, J. G., Boyce, W. E. and Siegmann, W., *Exploring Calculus with MAPLE*, Addison-Wesley, Reading, 1993

MacCallum M. A. H. and Wright F. J., *Algebraic Computing with REDUCE*, Oxford University Press, Oxford, 1991

Rayna G., *REDUCE Software for Algebraic Computation*, Pitman, London, 1983

Steeb W.-H., *Chaos and Fractals, Computations and Algorithms*, Bibliographisches Institut, Mannheim, 1992

Steeb W.-H., *Algorithms and Computation with Turbo C*, Bibliographisches Institut, Mannheim, 1992

Steeb W.-H. and D. Lewien, *Algorithms and Computation with REDUCE*, Bibliographisches Institut, Mannheim, 1992

Steeb W.-H., D. Lewien, O. Boine-Frankenheim, *Object-oriented programming in science with C++*, Bibliographisches Institut, Mannheim, 1993

Ueberberg, J., *Einführung in die Computeralgebra mit REDUCE*, Bibliographisches Institut, Mannheim, 1992

Wolfram, S., *Mathematica*, second edition, Addison-Wesley, Redwood City, 1991

# Index

| | |
|---|---|
| Angular momentum | 28, 30, 32 |
| Anharmonic oscillator | 24 |
| Anticommutation relation | 78 |
| Bose coherent states | 90 |
| Bose operators | 86, 88, 92 |
| Casimir operator | 30 |
| Chebyshev polynomials | 126, 178, 181 |
| Clebsch-Gordon series | 130 |
| Coherent states | 90 |
| Commutation relations | 22, 86 |
| Commutator | 14 |
| Conjugacy classes | 151 |
| Conservation law | 4 |
| C++ programs | 179 |
| Cumulant expansion | 170 |
| Delta function | 118 |
| Degeneracy of eigenvalues | 78 |
| Differential cross section | 106 |
| Dirac equation | 94 |
| Direct sum | 75 |
| Discrete Fourier transform | 146 |
| Dispersion relation | 94 |
| Driven two level system | 64 |
| Eigenvalue equation | 9 |
| Elastic scattering | 104 |
| Exceptional points | 108 |
| Fermi operators | 78, 82 |
| Fourier expansion | 148 |
| Fourier transform | 142, 146 |
| Free electron spin resonance | 66 |
| Gamma matrices | 94, 140, 175 |
| Gauge theory | 62 |
| Global gauge transformation | 62 |
| Gram-Schmidt orthogonalisation process | 158 |
| Group theory | 150 |
| Harmonic oscillator | 18, 20 |
| Heaviside function | 118 |
| Heisenberg equation of motion | 14 |
| Heisenberg model | 74 |

| | |
|---|---|
| Helium atom | 48 |
| Hermite differential equation | 19 |
| Hermite polynomials | 124 |
| Hubbard model | 82 |
| Hydrogen atom | 42, 44 |
| Hypergeometric functions | 132, 136 |
| Ising model | 70 |
| Korteweg de Vries equation | 160, 161 |
| Kronecker product | 70, 71, 72, 74, 76, 140 |
| Laguerre polynomial | 122 |
| Lax pair | 160 |
| Legendre polynomial | 120, 158, 159 |
| Leverrier's method | 172 |
| Lie algebra | 34 |
| Lippmann Schwinger integral equation | 167 |
| Local gauge transformation | 62 |
| Main quantum number | 45 |
| MAPLE | 3, 176, 177 |
| MATHEMATICA | 2, 174, 175 |
| Matrix representation | 10 |
| Pade approximation | 166 |
| Paraboloidal coordinates | 53 |
| Parseval's equation | 142 |
| Pauli spin matrices | 14 |
| Perturbation theory | 98 |
| Product ansatz | 8 |
| Pseudo-differential operators | 162 |
| Quantum groups | 154 |
| Quartic potential | 92 |
| Radial eigenvalue equation | 44 |
| Radial quantum number | 45 |
| Radial symmetric potential | 38 |
| REDUCE | 1 |
| Riccati differential equation | 54 |
| Scattering | 58, 104 |
| Schrodinger equation | 4 |
| Separation ansatz | 8 |
| Sign function | 118 |

| | |
|---|---|
| Soliton theory | 160, 164 |
| Sommerfeld fine structure constant | 42, 50 |
| Sommerfeld radiation condition | 104 |
| Spherical coordinates | 32 |
| Spherical harmonics | 128, 130 |
| Spin matrices | 88, 140 |
| Spin-1 Lie algebra | 36, 175, 177 |
| Stark effect | 52 |
| su(3)-Lie algebra | 34 |
| Trace | 172 |
| Trial function | 12 |
| Vacuum state | 78, 86 |
| Variational principle | 48 |
| Wave packet | 6 |
| WKB-solutions | 26 |
| Yang-Baxter equation | 154 |